Advanced Sciences and Technologies for Security Applications

The series Advanced Sciences and Technologies for Security Applications comprises interdisciplinary research covering the theory, foundations and domain-specific topics pertaining to security. Publications within the series are peer-reviewed monographs and edited works in the areas of:

- biological and chemical threat recognition and detection (e.g., biosensors, aerosols, forensics)
- crisis and disaster management
- terrorism
- cyber security and secure information systems (e.g., encryption, optical and photonic systems)
- traditional and non-traditional security
- energy, food and resource security
- economic security and securitization (including associated infrastructures)
- transnational crime
- human security and health security
- social, political and psychological aspects of security
- recognition and identification (e.g., optical imaging, biometrics, authentication and verification)
- smart surveillance systems
- applications of theoretical frameworks and methodologies (e.g., grounded theory, complexity, network sciences, modelling and simulation)

Together, the high-quality contributions to this series provide a cross-disciplinary overview of forefront research endeavours aiming to make the world a safer place.

The editors encourage prospective authors to correspond with them in advance of submitting a manuscript. Submission of manuscripts should be made to the Editor-in-Chief or one of the Editors.

More information about this series at http://www.springer.com/series/5540

Peter Lehr

Counter-Terrorism Technologies

A Critical Assessment

Springer

Peter Lehr
CSTPV, School of International Relations
University of St. Andrews
St Andrews, Fife, UK

ISSN 1613-5113 ISSN 2363-9466 (electronic)
Advanced Sciences and Technologies for Security Applications
ISBN 978-3-030-08132-4 ISBN 978-3-319-90924-0 (eBook)
https://doi.org/10.1007/978-3-319-90924-0

Printed on acid-free paper

This Springer imprint is published by the registered company Springer International Publishing AG part of Springer Nature.
The registered company address is: Gewerbestrasse 11, 6330 Cham, Switzerland

Contents

Chapter 1
Introduction: Terrorism as a Threat to Open Societies

Abstract In the introduction, I offer an overview of the book and the issues discussed in it. I introduce the individual chapters, but also familiarize readers with the basics of risk assessment and risk management as parts of the field of Critical Infrastructure Protection (CIP). In particular, I briefly talk about criticality, vulnerability, and threat, to then address questions on how to manage or mitigate risks posed by terrorism. In that regard, I highlight the role of technology in countering this threat and its relevance for both CIP and Critical Infrastructure Resilience (CIR). In a nutshell, the introduction aims at firmly anchoring the book in the field of CIP/CIR in order to give it a shelf life that goes beyond the current threat posed by Al Qaeda and ISIS/Daesh.

Keywords Critical Infrastructure Protection (CIP) · Risk assessment · Risk management · Agency of things · Future of terrorism

Against the backdrop of the persistent terrorist threat posed by the combined forces of Al Qaeda and ISIS (also known as IS, ISIL or Daesh) to western cities in the foreseeable future, I find myself giving an increasing number of presentations and talks that always seem to revolve around one central question: in how far can advanced counter-terrorism technologies contribute to securing or 'hardening' our cities, our public spaces, and the open societies of our liberal democracies against this threat? More often than not, the underlying assumption behind these questions seems to be that somehow, advanced technology could be a 'magical' bullet, or 'silver' bullet, in the fight against a wave of terrorism that now descends not only on our cities but on our very lifestyles with a ferocity not seen since the end of the Second World War. Surely, our advanced high-tech societies would be able to swiftly defeat these new 'urban' terrorists by unleashing the full power of our security technologies against them, even if in the short run at least we would have to trade in some of our civil liberties?

Obviously, there is a need for a critical appraisal of current and future counter-terrorism technologies – an appraisal that goes beyond a mere cataloguing of such technologies to critically assess their advantages and disadvantages as well as other potential pros and cons. In my view, it is not only a question of whether these

P. Lehr, *Counter-Terrorism Technologies*, Advanced Sciences and Technologies for Security Applications, https://doi.org/10.1007/978-3-319-90924-0_1

technologies work or not, or even what kind of short-term trade-offs regarding civil liberties their deployment might require. Rather, I am convinced that the uncritical use of these technologies, often indeed seen as 'silver bullets', may lead to unintended consequences in the shape of profound changes within our urban spaces, and also for our current concept of citizenship. Hence, in this book I will not only discuss the role of technology for defeating terrorism in general and with regard to securing our vulnerable urbanized and open societies in particular, but also their longer-term effect on our open societies and on our liberal democracies as such. However, before I start doing so, some broad introductory explanations are in order – first about the history of the 'how far can we go' question and the nature of the current threat, then about my theoretical approach that guides my argumentation.

1.1 Constructing Threats: Crime, Organized Crime, and Terrorism

To begin with the historical dimension, the 'how far can technology help us to combat this threat' question is not exactly a new one – it has been asked many times before with different kinds of threat in mind, probably ever since the first cities and towns emerged at some time during the 4th millennium BCE in Mesopotamia and China. And now this question is simply being asked again against the backdrop of rapidly changing societies driven by an exponential rate of technological progress on the one hand (Kurzweil 1999 calls this the 'law of accelerating returns'), and the emergence of a new threat on the other – the threat posed by global, transnational terrorism, whose agents also profit from the relentless march of technology. Very fittingly, Brazilian anthropologist Teresa Caldeira (2000, 1) calls this question-and-answer, action-reaction sequence the 'discourse of fear.' The 'fear' she writes about is the fear of violence – a violence that mainly manifests itself as violent crime in the Brazilian city of São Paulo that she focuses on. Terrorism as a special threat does not feature much in her work, which was finalized in 1998 and published in 2000, about a year before the watershed terrorist events of 9/11. Reading her book thus helps to understand that the claim 'the world's a different place now' after 9/11 is not necessarily true – at least not in the context of fear of violence, and actions, including technology-based actions, taken to respond to these fears. Caldeira argues as follows:

> In the last two decades [1980–2000], in cities as distinct as São Paulo, Los Angeles, Johannesburg, Buenos Aires, Budapest, Mexico City, and Miami, different social groups, especially from the upper classes, have used the fear of violence and crime to justify new techniques of exclusion and their withdrawal from traditional quarters of the cities. [...] The discourses of fear that simultaneously help to legitimize this withdrawal and to reproduce fear find different references. Frequently they are about crime, and especially violent crime. But they also incorporate racial and ethnic anxieties, class prejudices, and references to poor and marginalized groups. (Caldeira 2000, 1)

Now, about two decades later, we can easily substitute 'crime' with 'terrorism', while the rest of her argumentation remains virtually unchanged. Nowadays, the reference groups of 'racial and ethnic anxieties', of 'class prejudices', and of 'poor and marginalized groups' are, at least in Western Europe and in the United States of America, 'the Muslims' as such, including the current wave of refugees or migrants who mostly hail from Muslim countries of origin as well, such as Syria and Libya for example. What also remains unchanged is the 'constructedness' of the threat: indeed, if we simply substitute 'used the fear of violence and crime to justify new techniques of exclusion' with 'used the fear of terrorism', then we would be in agreement with some leading scholars from the side of Critical Terrorism Studies, for example John Mueller (2006) who convincingly argues that the threat of terrorism is over-blown.

Against the backdrop of ISIS-related terrorist attacks in Paris (13/14 November 2015, 137 killed and 368 injured), Brussels (22 March 2016, 32 killed and 316 injured) or Nice (14 July 2015, 86 killed and 434 injured), Berlin (19 December 2016, 12 killed and 56 injured), Westminster (22 March 2017, 5 killed, 49 injured), Manchester (22 May 2017, 22 killed, 116 injured) or Barcelona (17 August 2017, 14 killed, 118 injured), the claim that the threat of terrorism has been exaggerated may sound far-fetched at first glance – but when seen from a statistical point of view, it is still far more likely to get killed by a lightning strike, die of a heart attack or in a car accident or to drown in a bathtub, for that matter, than to become a victim of a terrorist attack. Pointing at statistics however is of cold comfort for the immediate victims of terrorism (those who died or were injured) and the indirect victims (their relatives and friends) as well. Also, the spectacular and very visible nature of acts of terrorism as 'propaganda by the deed' or as 'theatre of violence' renders them impossible to ignore – especially not in a time of post-modern media: nowadays, a terrorist incident immediately spreads all over platforms such as Facebook, Twitter, Periscope or YouTube, 'allowing' all those not directly affected, and not even near the scene, to vicariously experience the horror as well – which, of course, greatly adds to the atmosphere of fear.

In a sense, this means that terrorism should be seen not so much as an attack on our physical wellbeing (although for the immediate victims, it undoubtedly is) but as an attack on our psychological wellbeing (Wolfendale 2007, 80) and on our 'basal security' (Jones 2004, 15) – defined as the "unarticulated affective sense of safety and trust through which one (sometimes unconsciously) judges and assesses risks" (Wolfendale 2007, 81). After all, terrorism explicitly aims at creating an atmosphere of fear by (mainly) targeting civilians in order to coerce a political actor (more often than not the state itself) to react in one way or another – and one of the reactions anticipated and hoped for by terrorists is an over-reaction by the government of the day, driven not only by self-interest but also by the mass media and the population at large according to the 'politics of the latest outrage', as Malcolm Anderson (2008, 228) calls it. Seen from that perspective, not reacting to terrorism is not an option: something has to be done to stop this threat one way or another. Security expert Bruce Schneier explains why this actually has to be the case from the point of view of a politician:

> Imagine two politicians today. One of them preaches fear and draconian security measures. The other is someone like me, who tells people that terrorism is a negligible risk, that risk is part of life, and that while some security is necessary, we should mostly just refuse to be terrorised and get on with our lives. Fast-forward 10 years. If I'm right and there have been no more terrorist attacks, the fear preacher takes credit for keeping us safe. But if a terrorist attack has occurred, my government career is over. Even if the incidence of terrorism is as ridiculously low as it is today, there's no benefit for a politician to take my side of that gamble. (Schneier 2013)

Schneier also quotes former US President Bill Clinton, who stated in the light of the reactions to 9/11 that "[when] people feel uncertain, they'd rather have somebody who's strong and wrong than somebody who's weak and right" (Clinton 2002). As we know from Clinton's immediate successor George W Bush's 'Global War on Terrorism' or GWoT, one way to deal with the threat, and to demonstrate the strong and (possibly) wrong leadership Clinton referred to, is to securitize this threat, which means elevating it from a mainly criminal problem that could be tackled by police and the judiciary via the so-called 'Criminal Justice Model' to a fundamental challenge to our state, to the body politic, to the fabric of our society and to our very way of life – hence, a threat that warrants the use of the military via the 'War Model' (both models will be discussed in the next chapter). The advantage of this approach is that the fight can be carried to the terrorist enemies and their oversea bases, say, in Afghanistan, Iraq and Syria, or Somalia. Within our own borders however, the use of the military either is far less straightforward for a variety of reasons (also to be discussed in the next chapter) or not permitted due to constitutional constraints. At the home front, strengthening our criminal justice system, especially intelligence services and police, is a possible way forward. Militarizing our police forces by way of specially trained and equipped Police Paramilitary Units (PPUs) or Special Weapons and Tactics (SWAT) teams while passing ever stricter anti-terrorism laws is one option, while harnessing cutting edge technology as a potential 'magic bullet' against terrorism is another – which explains the focus of this book.

1.2 Bottomless Barrels and Moving Targets: The Need for Proper Risk Management

Regarding the terrorist threat, I already mentioned 9/11 (difficult not to do in any publication dealing with aspects of terrorism and counter-terrorism even after nearly twenty years after the events) and the more recent attacks on Nice, Paris, Brussels, Berlin, Manchester and Barcelona. I could also add London 7/7, the Madrid train bombings of March 2005, or the Boston Marathon bombings of 15 April 2013 as further Al Qaeda or ISIS-related attacks on Western cities and Western lifestyles. And of course, we should by no means forget the Provisional IRA's attacks targeting various parts of London, such as Bishopsgate (bombing, 24 April 1993), Canary Wharf (bombing, 9 February 1996), Downing Street 10 (mortar attack, 7 February

1991), and London Heathrow International Airport (multiple mortar attacks, 9–13 March 1994), while on the European continent, the mostly assassination-style ambushes of the so-called Fighting Communist Organizations (FCOs) in the shape of the Italian Red Brigades (Brigate Rosse), the German Red Army Faction (Rote Armee Fraktion, RAF), the Belgian Communist Combatant Cells (Cellules Communistes Combattantes) and the French Direct Action (Action Directe) could be mentioned as well, or the comparatively few but deadly bombing attacks of rather more nebulous extreme right wing terrorist groups, for example the Munich Oktoberfest bombing of 26 September 1980 (12 people killed, 211 injured) or the Bologna railway station bombing of 2 August 1980 (85 people killed, more than 200 injured) – the latter still being the fourth deadliest terrorist attack in Europe at the time of writing, only surpassed by the attacks of Nice, Paris and Madrid. I will return to some of them later on. What is more important in the present context is that the terrorist threat emanates from a variety of actors, and manifests itself in a variety of different tactics in a seemingly ever-changing terrorist threat landscape – which also include a number of empty threats and hoax calls, both from 'real' terrorists but also from 'imagined' ones, which means from individuals who for one reason or another think it is in their interest to mislead police forces, fire fighters and ambulance crews who, as first responders, rush to the scene.

Seen from a counter-terrorism perspective, it is obviously of utmost importance to properly manage these threats – which means we have to discern 'real' threats from bogus or 'imagined' ones. Tellingly, in a conversation with me a couple of years ago, a high-ranking US official opined that the need to protect our societies involved both 'moving targets' as well as 'bottomless barrels.' By 'moving targets' he meant the habit of groups such as Al Qaeda, and now ISIS as well, to threaten attacks against facilities or structures within our societies whose smooth functioning is required for the "continuous flow of goods and services" (PCCIP 1997) – facilities and structures that thus form part of what is known as critical infrastructure (CI).[1] These threats usually appear on websites and in chat groups, without necessarily resulting in any attacks – quite often, there is much Internet chatter about allegedly impending attacks that then never happen. Hence, not all of these threats are taken seriously by intelligence services or police forces, while some others are. As a consequence, security measures are stepped up for a while, until authorities are convinced that the threat either has passed or was bogus from the outset. Threats may be voiced against airlines and airports, ferry terminals, railway and bus stations, government buildings, sport arenas, nuclear power stations, shopping malls, or other parts of our critical infrastructures, forcing police and intelligence services to re-focus again and again – which is why the label 'moving targets' is very appropriate indeed. What is meant by 'bottomless barrels' is self-evident: employing and

[1]According to HSPD-7, Critical Infrastructure or CI can be defined as "those systems and assets, whether physical or virtual, so vital to the United States that the incapacity or destruction of such systems and assets would have a debilitating impact on security, national economic security, national public health or safety, or any combination of those matters." (HSPD-7, 17 December 2003). This definition also applies for any other modern state.

allocating security forces to guard a broad range of facilities and other assets belonging to critical infrastructure is very costly indeed. Hence, as can be imagined, having a clear and carefully drawn plan for what is called critical infrastructure protection (CIP) is a must: there will never be enough police officers or even soldiers available to protect all assets at the same time, and there will most definitely never be enough money available in the first place. Those tasked with protecting critical infrastructure have to make do with what they have and develop contingency plans on the basis of that. Thus, 'failing to plan' indeed is 'planning to fail'.

In the following chapters, the focus will be on urban guerrilla warfare as the most likely threat posed by terrorists of most persuasions including lone wolves to Western cities, and by extension, to Western open societies and the Western way of life in general. However, when seen from a critical infrastructure protection (CIP) perspective, 'threat'[2] is only one aspect of a much more complicated exercise that aims at assessing and mitigating the risk posed by terrorists to our critical infrastructures – 'risk' broadly understood in this context as "a function of consequences of a successful attack against an asset; and the likelihood of a successful attack against an asset" (API/NPRA 2004, 3).[3] Actually, it can be said that 'risk' is a function of 'criticality',[4] 'vulnerability'[5] and 'threat'[6] – which also means 'risk' and 'threat' are not the same thing – or 'vulnerability' and 'risk', for that matter. This is a crucial observation to make, because most of the worst-case scenarios developed by various parties tend to dwell on detected vulnerabilities which, as bottomless barrels in their own right, are "virtually unlimited, and would require extraordinary resources to address" (Decker 2001, 3). Brian Jenkins takes a similar point of view, arguing that "in vulnerability-based assessment, [terrorists'] actual capabilities, ambitions, and fantasies blur with our own speculation and fears to create what the terrorists want: an atmosphere of alarm" (Jenkins 2006, 25).

Quite obviously, projecting our own fears onto the terrorists no matter whether or not they have the capabilities and the inclination for carrying out our imagined worst-case scenarios usually results in us getting the threat wrong, and to the public and private sectors allocating scarce funds to protect their assets against largely non-existent dangers. Furthermore, it should also be emphasized that the factor 'vulnerability' is actually one we can, to a certain extent, eliminate or at least mitigate. After all, vulnerabilities are 'designed in' to many of our critical infrastructures via architecture, procedures, and policies, and thus can be 'designed out' as well (Hunt

[2]Threat is usually defined as the intention and capability of terrorists to carry out operations that can harm us.

[3]It should be mentioned at least in passing that CIP normally employs an 'all hazards' approach – i.e. a risk management plan that is valid for all kinds of risks, be they terrorist, criminal, human error, or natural disasters such as floods, earthquakes, blizzards etc. In this book however, we only deal with risked posed by terrorists.

[4]The key question according to Rosenzweig/Kochems (2005, 3) is [what] is the asset's function or mission and how significant is it?"

[5]Defined according to Roper 1999 as "a weakness that can be exploited to gain access to a given asset".

[6]Expressed as a formula, it would look as follows: $\mathbf{R = f(C,V,T)}$.

and Kellerman 2000, 3). A simple example of 'designing out' vulnerabilities would be substituting previously used conventional glass panes with more secure shatter-proof variants for public buildings such as airport or railway terminals: this would eliminate clouds of lethal shards scything through people in the case of a terrorist attack, or an earthquake for that matter. As I will explain later in this book, counter-terrorism technology plays an important role in this crucial process of 'designing out', as does the new urban architecture that aims at removing weak spots from the very beginning, that is in the 'blueprint' phase. But for the moment, it is important to repeat the argument made by Decker: in order to protect open societies against terrorist attacks in general, and terror attacks targeting Western cities in particular, a careful balance of the three elements criticality, vulnerability and threat is required – otherwise, a lot of time and effort will simply go to waste, and a false sense of security will be created.

Ironically, a very similar quick risk assessment as just explained also needs to be carried out by the terrorists themselves if they intend to really hurt us – and in order to minimize the effects of 'Murphy's Law (according to which everything that can go wrong will go wrong) and other issues usually accompanying highly complex operations. As Homer-Dixon (2009) opines, "terrorists must be clever to exploit [modern societies'] weaknesses. They must attack the right nodes in the right networks. If they don't, the damage will remain isolated and the overall network will be resilient." Homer-Dixon is confident that modern terrorists will master this art eventually. And as soon as they master the art of 'complex terrorism' as he calls it, modern terrorists will be able to reap havoc:

> Complex terrorism operates like jujitsu—it redirects the energies of our intricate societies against us. Once the basic logic of complex terrorism is understood (and the events of September 11 prove that terrorists are beginning to understand it), we can quickly identify dozens of relatively simple ways to bring modern, high-tech societies to their knees. (Homer-Dixon 2009)

This is why the protection of our critical infrastructures based on proper risk assessments is imperative – and more important than ever before. Hence, critical infrastructure protection and risk management will form one of the underlying theoretical themes of this book, influencing and shaping my discussions throughout its chapters.

1.3 'Agency of Things': Putting the 'Critical' into Critical Infrastructure Protection

Now it is time for a warning: the still relatively new art of Critical Infrastructure Protection (CIP) focuses – as it should – on risk assessment and risk management as just discussed. Managing the risk also includes risk mitigation and disaster relief, bringing succour to the immediate victims of such incidents, and in many cases even to the indirect ones, for example the relatives and friends of the immediate victims, and to the first responders who oftentimes have to deal with very gruesome scenes at the sites of the incidents. It is now well accepted that Post Traumatic Stress Disorder (PTSD) also affects the first aiders and helpers – not only soldiers returning from the battlefield. However, there are some questions which are not normally asked in the context of CIP. In my opinion, this makes CIP as a discipline vulnerable to some criticism – just like the so-called 'orthodox' terrorism studies and their underlying assumptions, which are challenged by 'critical' terrorism studies.

In the case of 'orthodox' terrorism studies, one of the most formidable challenges coming from the direction of 'critical' terrorism studies consists of the claim that the former are far too state-centric, in the sense that our own liberal democracies' involvement in terrorism tends to be ignored in favour of a focus on terrorist sub-state actors fighting against us. The underlying assumption of orthodox terrorism studies, so the 'critical' claim, is that our Western liberal democracies are only the victims of terrorism – but never the perpetrators. Hence, terrorism is what the 'others' do: sub-state actors fighting against us, obviously, while all those sub-state actors who fight for us presumably are noble 'freedom fighters.' Another often-heard challenge is that 'orthodox' terrorism studies largely ignore state-terrorism due to the exclusive focus on sub-state actors, even though terrorism can well be seen as a tactic that can be used by all kinds of actors – states included. Now, it is certainly true that there are, indeed, conservative think tanks and institutes who see it exactly as critical terrorism scholars claim. However, from my own experience as a scholar at the centre for the Study of Terrorism and Political Violence (CSTPV) at the University of St Andrews, one of the often mentioned alleged fortresses of 'orthodox' terrorism studies, I can say that the truth usually is far more complicated: in our case, the opinions range from the left-leaning to the conservative, and trying to arrive at a common opinion or 'official' CSTPV comment to an event is a bit like 'mission impossible': the idea of 'herding cats' readily comes to mind.

When it comes to CIP studies, there also seems to be a set of underlying assumptions which remain unchallenged – not necessarily due to ill intent but rather to the mostly technocratic nature of the many scholarly articles and the relatively few monographs on this subject that still is in its infancy: just as in the case of only slightly older terrorism studies, the bulk of research has been undertaken in the post-9/11 environment. Against the backdrop of the relatively short history of CIP research, it is thus not that surprising that the focus still lies on the 'how to make it work' or 'how to improve it' questions, and on issues relating to risk assessment and risk management – the only 'critical' part usually amounting to just a "critical analysis of the remaining gaps" (Collins and Baggett 2009, 2) in the array of protective measures taken. At times, these rather mundane matters of a mainly technical and sometimes mathematical nature are discussed in conjunction with a more insightful debate on whether it would be better to focus on critical infrastructure <u>protection</u> (CIP) or on critical infrastructure <u>resilience</u> (CIR) – the latter dealing with matters such as survivability of infrastructures and of business continuity after an attack occurred. In itself, this debate is indeed quite interesting, but it is of less interest to us in the present context since our focus clearly is on protection as well.

What is missing in the bulk of these contributions is the crucial question of where a purely technocratic CIP ultimately leads: surely, hardening our cities against terrorism by way of deploying security-related technologies will have an impact on them and on our daily lives? Don't these technologies influence our behaviour, and our day-to-day lives as well? Cultural anthropologists are familiar with the term 'agency of things.' 'Agency of things' means that it is not only people who have agency, defined as the capacity of an actor to act in, or influence a, given environment. Objects, or 'things', can also wield such an influence. As Latour puts it, agency can be attributed to "anything that modifies a state of affairs by making a difference in an actor" (Latour 2005, 71). Examples for this abound, probably starting with the invention of the wheel or the discovery of fire and ranging from the very mundane to the highly complex, so one will suffice here: just consider the changes brought to people's lives via information technology, especially the Internet. It is actually rather difficult to explain to members of the younger generation how the previous ones managed without it. I remember from my own early-career days in the 1990s as a lecturer in South Asian Politics that my colleagues and I had to wait for a couple of days or even weeks to get the newest edition of the Times of India and the Hindustan Times, arriving as hard copies via (expensive) air mail or (much cheaper but far slower) sea/surface mail. Nowadays, if I still feel so inclined, I can read both newspapers online at the very day they appear – and lots of other newspapers from all over the world as well. What my colleagues and I found out very quickly however was that this now sheer limitless amount of information at our fingertips came with a certain price tag: more time for reading equals less time for thinking – it is as simple as that. And as most of us also learned very quickly, there always seems to be one more link to click, and one more article to read. In short, from the end of the 1990s onwards, the Internet changed my working habits, and those of (most of) my colleagues. And this is exactly what the 'agency of things' refers to.

In the case of CIP, it is quite obvious that the technologies developed and deployed in order to protect us also have agency as just defined and explained: they certainly 'modify a state of affairs by making a difference in an actor.' First of all, they are explicitly meant to modify a state of affairs by hardening our vulnerable critical infrastructures in a risk assessment/management/mitigation process. Among these measures taken is the deployment of security technology, such as chain-link fences, CCTV systems, access controls, bollards and others – all of which will be discussed in the second part. These measures can have an immediate effect on our daily behaviour: for example, if a previously openly accessible street is closed to the public due to heightened security measures, then we might have to take another route to our destination. In the case of airports, we might have to drop off travelling family members or friends at a distant drop-off point, and not directly at the terminal any longer, due to the security barriers in place to prevent car or truck bombing attacks. To access certain shopping malls, plazas, or office or residential buildings, we might now be required to go through a metal scanner and/or have our bags and briefcases searched even though, until recently, we simply walked in. When it comes to access control, woe betide those who tend to forget their swipe cards: the door remains closed – which can be quite awkward if it's the weekend, and the swipe card together with ones' keys, wallet and mobile phone are sitting on the desk, now firmly out of reach at the wrong side of the access-controlled door when one returns from a short break. Getting used to wear lanyards with the 'magic' card fixed to it hence is a good idea. All these are examples of modified behaviour, and of the agency of things.

It is precisely this modification of behaviour by way of the agency of things that I find interesting, but also quite underexplored as it stands in current debates about CIP processes and procedures. If this issue is touched upon at all, then only with regard to a modification of behaviour of the 'other' in the shape of terrorists or violent criminals. After all, risk management and risk mitigation aim at forcing terrorists and criminals to desist by way of hardening vulnerable parts of our critical infrastructures in general and our cities in particular, as already discussed. Seen against this particular backdrop, the basic and unchallenged assumption seems to be that by making our societies, our cities, and by extension us, safer from the threat posed by terrorists and violent criminals, there cannot possibly be any negative side effect worthwhile commenting on. The very idea that the silver bullet might also hit us, at least potentially and as a ricochet, seems to be so far-fetched and outlandish that it usually does not even come to the mind of those researching the art of critical infrastructure protection.

Even in the rare cases that do acknowledge that a proportionate balance must be found between security requirements and people's lives including their privacy, this issue is not really followed up – mentioning it once seems to be enough, as if this would magically make it go away. In the United Kingdom, the so-called Communications Data Bill as part of the larger Investigatory Powers Act demonstrates that this is not necessarily the case. Better known as *Snooper's Charter,* this bill was passed by the British Parliament on 16 November 2016. It includes the passage that "[service] providers must now store details of everything you do for 12 months – and make it accessible to dozens of public authorities" (Carlo 2016). Whether this charter still should be seen as striking a proportionate balance between security and privacy probably lies in the eyes of the beholder, but I am sceptical. One could of course also argue from a utilitarian perspective that domestic security interests outweigh individual privacy concerns – after all, the freedom of fear of terrorism and violent crime for the many might well trump the fear of loss of privacy by a few. Again, I am sceptical. Be that as it may, it is obvious that a discussion of CIP-related security and surveillance technologies also require critical scrutiny with regard to their impact on the society they promise to protect, and on that society's civil liberties. Critically scrutinizing their impact on society is exactly what I intend to do as well in this book, more precisely in its third part.

1.4 Predicting Terrorism of the Future – A Caveat

Since this book is not only about the present but also about the foreseeable future, it might be appropriate to add some disclaimers. To begin with, a lot of pen has been put to paper concerning our ability, or rather, inability, to predict future events. Amongst the many explanations offered for why even experts oftentimes get it wrong is the one of so-called 'black swan' events that, seemingly out of the blue, interfere with the expected smooth flow of history – largely unpredictable events in other words, such as the devastating terrorist attacks of September 11, 2001. Such 'black swan' events render predicting the future on the basis of current developments notoriously difficult. This implies that this book deals with just one possible future concerning terrorism and counter-terrorism. For example, we could also look at cyber terrorism as the future wave of terrorism – as many terrorist experts actually do. Indeed, it would be good to look at scenarios in which currently active hackers, or future ones, could go 'political', i.e. turning from various non-political (but possibly criminal) forms of hacking – white hat, grey hat, or black hat – to politically motivated forms of hacking, including cyber terrorism proper.[7] Against the backdrop of long-existing expertise from groups like Anonymous, Lulzsec and the Chinese Inj3ct0r Team, the suggestion that cyber terrorism might happen in the future is a valid one. However, including Critical Information Infrastructure Protection or CIIP in this book as well would have necessitated a second volume since this is an art in itself.

Another possibility that should not be discarded without further thought would be a proliferation of so-called *lone wolf* terrorists, either acting on their own, or on behalf of existing ideologies and movements for which they carry out 'enabled' or 'remote-controlled' attacks (to use the proper counter-terrorism terminology). In this context, it should be noted that emerging technologies such as 3-D printers or readily available bio kits could make planning and carrying out terrorist attacks easier than ever before, while potentially also rendering them ever more lethal. If we take a look at the 'Alphabet bomber' Muharem Kurbegovic for example, it is evident that similar plans involving chemical agents would be much easier to pursue in our times than in the 1970s when he hatched his plans. The Internet now offers all the information in a matter of hours of concentrated research that Kurbegovic had to dig out in libraries and archives over several months. In a sense, these are the possible 'black swan' events that are somewhat predictable – which means they probably should better be called 'grey swans' or, to paraphrase Donald Rumsfeld, 'known unknowns.' As such, these 'enabled' or 'remote-controlled' attacks can, to a certain extent at least, be kept in mind as a new high-tech tactic that requires an equally high-tech response.

---

[7] According to Nelson et al. (1999, 9), cyberterrorism can be defined as "the unlawful destruction or disruption of digital property to intimidate or coerce governments or societies in the pursuit of goals that are political, religious or ideological." Dorothy Denning adds to that another crucial criterion: "Further, to qualify as cyberterrorism, an attack should result in violence against persons or property, or at least cause enough harm to generate fear" (Denning 2000).

Yet another caveat is in order in the present context: counter-terrorist forces are too focused not only on the current threat but also on the current set of perpetrators and their causes – Al Qaeda and ISIS on the one hand, and all those so-called 'home-grown' terrorists fighting on their behalf on the other. Of course, this is done for a perfectly good reason: both movements constitute a 'clear and present danger', to put it in US American parlance. And so do all those jihadists of Western extraction who chose to join the civil war in Iraq and Syria and fight on behalf of ISIS – to then return and continue the fight in our midst: so far, about 30 percent of the foreign fighters from the EU (between 3922 and 4294 according to the International Centre for Counter-Terrorism in The Hague) returned to their home countries. Even though the bulk of them probably won't carry on with the jihad for a variety of reasons, a substantial number of well-trained and experienced fighters might be willing to do just that, especially after the loss of their strongholds in Raqqa/Syria and Mosul/Iraq. These individuals plus those who self-radicalize themselves without ever travelling to Iraq or Syria constitute the fringe group of the movements influenced by Al Qaeda and ISIS usually labelled 'home grown terrorists.'

This rather convenient label however is not unproblematic since it points at the end product of a potentially long radicalization process. Hence, Graeme Pearson, former senior police officer and ex-Director General of the Scottish Crime and Drug Enforcement Agency, firmly rejects it, stating that instead of calling them home-grown extremists, it would be better to think of them as "disaffected people in our communities" or "people who might be unbalanced in their view of what society is about" (as quoted in Swindon 2016). There is a reason why the term 'disaffected people' should be used instead of the more common, and far more specific, term 'home-grown extremists': by doing so, attention can be drawn on the earlier and mostly still unspecific stage of a gradual radicalization process that may lead such individuals towards extremist Islamism, but also to Right-Wing or Left-Wing extremism, or other manifestations of political violence. Arguably, there always have been, and there probably always will be, disaffected people in our communities who may well latch on to the cause celebre of the day, first of all in order to express their discontent but also because it gives them a new meaning for their lives. A fitting example from the past would be the famous *International Brigade* coming to the rescue of the Spanish republic fighting against General Franco's fascism in the 1930s. Many *brigadistas* hailed from the United Kingdom, by the way, and most of them eventually returned to their homes. In the 1960s and 1970s, it was the allure of Marxism-Communism that caught the imagination of disenfranchised youths in several Western European states – mainly in Italy where university graduates were confronted by a rather bleak future, but also in West Germany, especially in West-Berlin, and in France and Belgium. Likewise, for young Catholics living in Northern Ireland in the times of 'The Troubles', joining the IRA was at least an option to be considered. At the moment, the 'fads of the day' in most parts of Western Europe are Salafism-Jihadism and Right-Wing extremism, both appealing to different sets of disenfranchised youths. In Greece and, to a lesser degree, in France, a resurgence of (Neo-) Anarchism can be reported as well. Again, what the future brings cannot be predicted with any degree of certainty – but it is more than obvious that new ideologies or 'causes' appealing to disenfranchised parts of the population will spawn new waves of political violence including terrorism. Hence, it is important for the purpose of discussing the future of terrorism as well as of counter-terrorism to keep an open mind. Focussing on those parts of the population who feel disenfranchised for one reason or another is key in this regard.

1.5 The Road Ahead: Overview

To sum it up, this book will not only critically discuss the role of technology for defeating terrorism in general and with regard to securing our vulnerable urbanized and open societies in particular, but also their longer-term effect on our open societies and on our liberal democracies as such. In order to accomplish this, it is organized into three parts: the first focusing on threats and responses, the second examining counter-terrorism technologies, and the third commenting on the effects they might have on our cities, and our understanding of citizenship. The first part opens with a chapter that familiarizes readers with the basics of risk assessment and risk management as parts of the discipline of Critical Infrastructure Protection (CIP). In a nutshell, this first chapter aims at firmly anchoring the book in the field of CIP in order to give it a shelf life that goes beyond the current threat posed by Al Qaeda and ISIS.

On the basis of the arguments made in the first chapter, the following one then critically discusses current and foreseeable tactics and weapons used by movements such as Al Qaeda and ISIS – weapons that may well include chemical, biological, radiological and potentially even nuclear (CBRN) devices. Whether such devices will be used or not in the foreseeable future is, of course, debatable, and not all experts are convinced. Nevertheless, our highly vulnerable open societies are open in that regard as well: a successful strike with CBRN devices could be devastating – both with regard to human costs as well as with financial and economic costs. For example, a successful strike with a radiological dispersal device or RDD (better known as 'dirty bomb') against the City of London or Wall Street as strategic key assets critical for our modern western economies would close down affected parts for a long time due to radiological contamination. It goes without saying that it is not only a manifest terrorist threat that we need to fear: random acts of 'Mother Nature' also can be quite devastating. Recent incidents such as Hurricane Katrina or the blizzards affecting the US American east coast in the last years for example, are a telling reminder that modern urban society is only three meals away from anarchy. The broader themes of vulnerability of our open societies, and the criticality of certain parts of urban infrastructures, the existence and smooth functioning of which we tend to take for granted, will form the backdrop for the discussion of the growing terrorist threat to our cities, moving from assassination-style attacks of, for example, the German Red Army Faction (RAF) and the bombing attacks of the IRA in the early 1990s against the City of London to mass-casualty attacks targeting our Western way of life in general and our 'sinful' cities in particular by actors associated with Al Qaeda and ISIS. My main argument here is that our growing urban sprawls now provide terrorists with the 'urban jungle' that Carlos Marighella (1969) had in mind when he wrote his famous *Minimanual of the Urban Guerrilla* in the late 1960s.

As a consequence, the new wave of urban guerrillas needs to be neutralized as far as possible by appropriate counter-measures by police as well as by paramilitary and military forces who also have to move into cities in order to defend them – hopefully short of a tactic that was described as 'it became necessary to destroy the town in order to save it' (a US Army major to Peter Arnett on the destruction of the South-Vietnamese provincial town Bên Tre in February 1968). This new 'military urbanism' will be the focus of the second chapter. Drawing on riots in several French banlieues (suburbs) and on the London 2011 riots, I will however also point out that terrorists only form the tip of an iceberg of politically motivated violence triggered by an outbreak of latent conflicts in huge urban sprawls. Hence, I will conclude this chapter with the argument that under certain occasions, the Western urban space might not only turn into a contested space but even in a veritable battle space itself – the fate of cities such as Beirut, Bagdad, Grosny or Mogadishu can serve as telling examples in this regard. In the concluding fourth chapter, I will briefly inspect the possible consequences of terrorist *actions* and police/paramilitary/military *reactions* for our cities – and, of course, for us in general. Assuming that none of us is overly keen to have our own cities turned into veritable battle spaces as well, I will argue that against this backdrop, it is only logical to harness modern technology because it might give us an edge in this fight. After all, technology indeed might be just the magic or 'silver' bullet we are looking for.

The second part of this book will home in on such counter-terrorism technologies – both technologies already available or in development that promise an increase in safety and security when it comes to the dangers posed by these terrorists.[8] This part thus offers a detailed and critical discussion of such technologies, starting with intelligence-gathering, monitoring and surveillance technologies, ranging from CCTV/Smart CCTV to vehicle tracking to the use of biometrics such as fingerprints, iris scans, gait recognition and others to real-time data mining and 'big-data' analytics. I will touch upon Bentham's Panopticum, and also on some utopian dreams as expressed for example in the Hollywood blockbuster *Minority Report.* And of course, controversies around certain technologies such as early types of full-body scanners deployed at airports (initially also dubbed 'naked scanners' for good reason) or surveillance drones will also have to be included. Critically discussing these technologies will logically lead to an assessment of their current and future application as additions to our modern walls around infrastructure we deem to be critical, such as airports, certain government buildings, some residential areas (gated communities and closed condominiums) or whole parts of cities for that matter. The 'wall of steel', followed by the 'wall of glass' around the City of London offers a very instructive example here. Mentioning the City of London will

[8] It should be noted that I will solely discuss passive counter-terrorism technologies – active defences such as new and non-lethal weapons systems will not be discussed in this book.

also serve as a reminder that modern efforts to secure our cities did not start in the aftermath of 9/11 only – even if in that case, these measures were clearly driven by the fear of terrorism, and not of violent crime in general.

In the following chapter, I will continue the discussion of modern city walls to turn to matters of urban architecture and the 'new urban geography'. As already mentioned, vulnerability is the only element of the criticality-vulnerability-threat triad over which we have some control: to a certain extent, we can 'design out' many of these vulnerabilities, for example by arming our streets with sturdy types of street furniture such as concrete flower pots or benches made of steel that can double up as barricades against car or truck bombs, and by using shatter-proof glass – or by avoiding vulnerable lower-floor glass facades altogether for that matter. The new *One World Trade Center* in New York is a telling case in point here: it features a reinforced concrete base, extra-wide pressurized stairwells, and biological/chemical air filters on top of that. Furthermore, the interior is constantly monitored by about 400 smart CCTV systems featuring video-analytical software tailor-made to detect potential threats. Other topics to be included here are barricades and citadels (e.g. gated communities or access-controlled malls/public spaces) as "real and symbolic barriers […] that combine to bring an environment under the control of its residents" by transforming it into a "defensible space" (Newman 1972, 3).

In the third, analytical part, I will take the critical discussion further by arguing that the uncritical deployment of new surveillance and control technologies in parallel with the on-going outsourcing and privatisation of key services of the state could well lead to dystopias as envisaged in a rather prescient way by the so-called *cyperpunk* novels of the 1980s. The preceding section concluded with a discussion of measures taken to 'harden' our cities against terrorist and related threats. In this third section, I change the perspective to re-examine these measures from a different angle – not from a counter-terrorism one, but from the perspective of citizens living in such hardened cities in order to draw attention on the broader consequences of our quest for more security and the deployment of counter-terrorism technology. After all, barricades and citadels will have a measurable effect on the life in cities affected by these measures: certain spaces that used to be freely accessible to everybody will now be accessible only after thorough security checks, or only for individuals in possession of proper credentials (access cards, IDs…), or not at all. Certain semi-public spaces such as some suburban parts of cities will be out of reach for everybody but those who are living there. Again, this is nothing new that came into being only after 9/11 – the City of London, or the inner city of Belfast for that matter, went to a process of barricading and citadelization in the 1990s (London) and the 1970s (Belfast). However, the fact that some parts of the city that used to be freely accessible are no longer so, necessitating long detours, will become the norm in the foreseeable future. This will be made worse by the possibility that parts of the city might end up as 'no go' areas where one either does not venture at all or where one is well advised not to stop at traffic lights for fear of being robbed or carjacked – which was the case for certain parts of Brussels even prior to 9/11, by the way.

These changes in urban geography as a consequence of the deployment of counter-terrorism technology however are only the tip of an iceberg of changes. Hardening and securing open societies in general and Western cities in particular does not come cheap – which raises the question on how to pay for it. In a time of shrinking state resources by ever-growing public debts, an easy and already routinely chosen solution is outsourcing – a strategy that also sits well with conservative parties lobbying for a weak state in favour of the private sector and the 'free market.' At the time of writing, services that used to be seen as core services of the state have already been privatized, being run by private security companies (PSCs) now. Examples are certain forensic services, formerly run by the police, a growing number of prisons, or even policing and patrolling of certain areas, for example shopping malls or low-security military installations such as barracks – the latter until a couple of years ago guarded by the soldiers occupying the installation in question themselves, but now protected by private security companies as well. The ongoing privatisation of further state services also starts to affect other first responders, such as fire fighters and ambulance services. For example, in Calvados/France, the fire brigade already expects to be paid for responding to distress calls by elderly people or OAPs who had the misfortune to fall in their own apartments and were unable to get up on their own.

This 'pay-as-you-panic' approach (so the headline of an *Independent* article) is bound to spread in the foreseeable future – which reminds me of certain dystopias envisaged in prescient cyberpunk novels of the 1980s, such as William Gibson's *Neuromancer* or Neal Stephenson's *Snow Crash*: citizens will get exactly the level of the safety and security they pay for – not only with regard to their choice of the place they intend to live but also with regard to the choice of their private first responder, given the fact that the remaining understaffed state first responders might be unable to respond in a timely manner. I will leave it open to the reader to decide whether he/she is willing to follow my line of argumentation here, but I will provide enough examples to make readers pause and think at least. It is of course highly questionable whether science fiction written in the 1980s and 1990s should or could be seen as a blueprint for our 'real' futures – which is why I also take a look at current developments outside of our Western comfort zones to see how security technology and security-driven urban development shapes up their perceived security, and what impact it has on their lives. After all, these 'real world' developments might be better and more reliable predictors of our western future than cyberpunk novels. After having hopefully unsettled my readers talking about 'architectures of fear' and 'archipelagos of fear', I will then argue in conclusion to this section that all taken together, all the security technologies discussed here will have yet another side-effect that cannot possibly be easily discarded: it divides citizens into 'us' and 'them', hence into trusted ones and the (probably much more numerous) 'rest.' The trusted ones will breeze through security checks and walk or drive along corridors and streets earmarked for them, while the 'rest' will either be required to undergo time-consuming checks, or simply denied access. A couple of examples drawn from various Western and non-Western cities will illustrate these arguments.

In the (short) conclusion of this book, I will lead all the various threads together, summarize the main findings and arguments, and culminate in the (in my opinion) all-important question of how far can we, as Western liberal democracies, go in our fight against terrorism? Is everything allowed in our current quest for one hundred percent security, or are there certain things that we should leave well alone? I will remind readers of Benjamin Franklin's remark that 'those who would give up essential Liberty to purchase a little temporary Safety deserve neither Liberty nor Safety.' Whatever readers think about the foreseeable future in a time of international terrorism, they should be aware that the adoption of modern technology that ostentatiously aims at increasing our safety and security does not come free of cost: rather, these technologies, if adopted in an uncritical way and mistaken for 'silver bullets', could profoundly change our cities, our public spaces, and our civil society as well.

Finally, I should also mention that I wrote the three parts of the book in a way that they stand for themselves, and that they can be understood even without having read the other parts. Which means, if readers prefer to start with the second part on technology to then turn to urban terrorism and counter terrorism, they should feel free to do so. With that, it is time now to get into medias res.

Part I
Barbarians at the Gates: 'Sinful' Cities

To provide the proper background to the application of technology for the purpose of countering terrorism, in this first part of the book I will examine the action-reaction dynamics between terrorists (mainly but not exclusively those belonging in the orbit of Al Qaeda and ISIS) and counter-terrorists (intelligence agencies, police, paramilitary, military, judiciary). I will also establish the current and foreseeable threat posed to our open societies in general and our cities in particular. Basically, I will argue that the terrorist threat has evolved during the last decades, and that modern terrorists now can inflict greater damage than ever before not only on our open societies and cities but also on what is known as critical infrastructure – damage that can bring our modern, high-tech societies at least to a temporary stand-still. I will also argue that while our counter-terrorist forces are nowadays well prepared to react to this threat, 'reacting' as such might not be good enough since this costs valuable time – valuable time the terrorists can use to keep on killing, or to barricade themselves in suitable locations, using their hostages as human shields. The arguments developed in this part thus prepare the ground for the discussion of counter-terrorism technology – often seen as a possible 'magical' or 'silver bullet.'

Chapter 2
Actions: The Return of Urban Guerrillas

Abstract In this opening chapter of the first part of this book, I analyse the current threat of terrorism as regards the safety and security of our cities and our open societies. The themes of vulnerability of these open societies, and the criticality of certain parts of urban infrastructures the existence and smooth functioning of which we tend to take for granted form the backdrop to the discussion of the growing threat to our cities, moving from assassination-style attacks of, for example, the German Red Army Faction and the bombing attacks of the IRA against the City of London to mass-casualty attacks targeting our Western way of life and our 'sinful' cities as such by actors associated with Al Qaeda and ISIS/Daesh on the one hand and attacks by way of weaponizing 'mundane objects' such as cars, vans, trucks or simple knifes for that matter. My main argument here is that our growing urban sprawls now provide terrorists with an 'urban jungle' Marighella could only dream of when he wrote his Minimanual of the Urban Guerrilla in the late 1960s.

Keywords Urban guerrillas · Terrorists' weapons and tactics · Propaganda by the deed · Complex terrorism · Conventional Low Impact (CLI) tactics · Conventional High Impact (CHI) tactics · Unconventional High Impact (UHI) tactics · Weapons of Mass Destruction (WMD) · Marauding attacks

When it comes to terrorist actions in the 'here and now' (at least as regards Western liberal democracies), it is a wise idea to re-examine Carlos Marighella's famous *Minimanual of the Urban Guerrilla* (Marighella 1969) and the underlying concept of 'urban guerrilla'. Regarding the concept of 'urban guerrillas' itself, Anthony Burton defines them as individuals who are fighting 'a little war' as this is what the term 'guerrilla' means. As he further explains, these urban guerrillas are basically the same as the 'classical' rural guerrillas from which they depart "only in their use of the slum and the ghetto instead of or in addition to the jungle and the peasant village" (Burton 1975, 10). Although the idea of urban guerrilla warfare, especially in Western cities, took a long time to translate from a theoretical concept into a practical one, it seems to have finally made this transition. The events in Mumbai 2008, Paris 2015 and Brussels 2016 are most indicative here, although it can be argued that the two sieges or hostage crises of Moscow Nord-Ost (October 2002) and

P. Lehr, *Counter-Terrorism Technologies*, Advanced Sciences and Technologies for Security Applications, https://doi.org/10.1007/978-3-319-90924-0_2

Bezlan (September 2004) already offered a first glimpse at the shape of things to come. And what we know about the motivations as well as the weapons and tactics of terrorist movements such as Al Qaeda and ISIS makes it very likely that such attacks will continue for the foreseeable future. The reasoning behind this assumption is a simple one. As John Robb (2007, 57) states, "[attacks] that work will be repeated until they no longer succeed, and then their perpetrators will try out various other tactics until they find a new success. It's that simple." Indeed, experience shows that quite often, terrorist organisations are copycats: they observe what other groups do, assess whether it worked, and if yes, whether it would work for them as well. If the answer is positive, they might try to add the tactic and/or weapons system to their own list of capabilities. Examples would be the rapid proliferation of politically motivated airliner hijackings in the 1960s and 1970s, and of the equally quick adoption of suicide bombings from the 1980s onwards. Hence, it is entirely plausible to argue that the successful (from the terrorists' perspective) so-called 'marauding attacks' – that is, combined shootings and bombings – of Mumbai, Paris and Brussels will also enter the repertoire of various terrorist groups, including those not belonging to the Al Qaeda-ISIS ideological environment. This kind of 'institutional learning' from one another is far more plausible than the sudden development of something completely new and untested (on institutional learning, see for example Dolnik 2009). Black swans, defined as unexpected and unforeseeable events, do exist as we know, but still, they are few and far between. Which also means our best bet for counter-terrorism is to take a close look at those tactics that are most likely to be used, and not at the more improbable scenarios – which is what I shall do now.

2.1 Propaganda by the Deed in Action: Terrorists' Weapons and Tactics

It is tempting to begin with an overview of weapons systems most frequently used by terrorist groups, but for brevity's sake, I shall focus on the most frequent tactics instead. These can be broken down into three different categories. The first category is that of conventional low impact or CLI attacks that are about, as Brian Jenkins (2006, 118) famously put it, a lot of people watching but not a lot of people dying. Apart from non-lethal tactics such as sabotage or arson, CLI attacks also include assassination, kidnapping, and hostage taking. The second category of conventional high impact or CHI attacks usually revolves around a lot of people watching, and a lot of people dying as well. Hence, its tactics include shooting sprees, sieges (large-scale hostage crises), attacks on transport systems such as planes, trains, busses or ships, car and truck bombings as well as suicide bombings, the use of captured airliners as flying bombs or, potentially, the use of ships as floating bombs. The third category of unconventional high impact or UHI attacks includes weapons of mass destruction (WMDs) such as chemical, biological, radiological and nuclear (CBRN) devices. Cyber terrorism also is frequently mentioned in the context of UHI scenarios. However, unlike CLI and CHI attacks, UHI events so far occurred only rarely and without as much success as the perpetrators presumably hoped and counter-terrorists feared.

The history of CLI terrorism can be traced back to classical antiquity at least and the anti-Roman Sicarii ('dagger men' from 'sicca' = dagger) active in Judea in the first century CE and the Shiite/Nizari Ismaili sect of the Assassins active in the Near and Middle East in the twelfth and thirteenth century CE who fought against various Muslim rulers as well as against the rulers of some Christian Crusader states that had been established in the wake of the First Crusade (Lewis 1967). Both movements used the same tactic: their members assassinated high-ranking officials, stabbing them to death in the open, preferably in public places. Their weapon of choice was the dagger – a weapon of choice that, in the shape of the knife, recently made an unexpected comeback within terrorism. The anarchist terrorists of the second half of the nineteenth century and the first decades of the twentieth century already had more advanced means at their disposal than such close-quarter weapons to carry out very similar acts of CLI terrorism: pistols and bombs. For example, US presidents Abraham Lincoln (assassinated 15 April 1865), James A. Garfield (assassinated 19 September 1881) and William McKinley (assassinated 6 September 1901) were shot dead or fatally injured with pistols, while Russian Tsar Alexander II (assassinated 13 March 1881) was killed by a crude grenade thrown at him – after he miraculously survived the first grenade attack just seconds earlier.

The central idea behind these CLI activities was twofold: a) tempting the state under attack to over-react, and b) to trigger an uprising of the population in general against the state. Leading Anarchist thinkers such Paul Brousse, Carlos Pisacane, Mikhail Bakunin or Johann Most called this approach the 'propaganda by the deed' (Fleming 1980). Since winning the support of the population as such (or at least parts thereof) implied the avoidance of what nowadays is known as 'collateral damage', i.e. the accidental killing of innocent bystanders in order not to alienate potential supporters, assassination-style attacks were a logical choice for these early terrorist movements. The so-called *New Left* wave of terrorism emerging in the 1960s in Western Europe, the United States and Japan had similar immediate objectives. Taking up the fight against what they perceived to be 'capitalist-imperialist' oppressor states roughly five decades after the anarchist wave expired, they had access to a variety of more sophisticated weapons than the anarchists. Submachine guns, assault rifles, high-velocity sniper rifles or rocket-propelled grenades greatly facilitate acts of terrorism – which is why Bjørgo (2005, 3) calls modern weapons technology a *facilitator* or *accelerator causes* that makes "terrorism possible or attractive." However, as for example the 14 July 2016 Nice Massacre demonstrated, the availability of more modern, more accurate and more powerful weapons should not be over-stated: in this incident, the weapon of choice was a 20-tonne refrigerated truck, used simply but very effectively to mow down as many civilians as possible. The lone-wolf perpetrator, Mohamed Lahouaiej Bouhlel, apparently followed the advice of ISIS spokesperson Abu Muhammad al-Adnani al-Shami who, in

September 2014, suggested to all the movement's potential supporters to use any means to their disposal if they had no access to explosives or weapons:

> If you are not able to find an IED [Improvised Explosive Device] or a bullet, then single out the disbelieving American, Frenchman or any of their allies. Smash his head with a rock, or slaughter him with a knife, or run him over with your car, or throw him from a high place, or choke him, or poison him. [...] If you are unable to do so, then burn his home, car, or business. Or destroy his crops... If you are unable to do so, then spit in his face (as quoted in Lister 2015).

That 85 citizens died either immediately during the attack or later in hospital (the last victim passed away on 4 August), and that more than 200 were seriously injured, is as chilling as it is instructive for counter-terrorism – but I shall return to the broader implications of this incident later. In the present context, suffice it to say that the availability of modern weapons should not be over-emphasized as a facilitator or accelerator cause for modern terrorism. Rather, for terrorist groups active after the 2nd World War, a far more important element of their struggle was (and still is) a ready access to modern global mass media that works for them as a useful impact magnifier by offering them the 'oxygen of publicity', as British Prime Minister Margaret Thatcher once famously put it. Nowadays, outlets such as print media, radio stations, and (since the 1980s) 24/7 global TV networks and their army of correspondents cater for their worldwide, media-hungry customers with an insatiable thirst for spectacular stories, since the early 2000s supplemented by social media platforms such as YouTube, Facebook or Twitter. Thus, much more than the anarchists could ever hope for, terrorists from the 1960s onwards found a potentially global audience, informed, entertained and served by an ever-growing number of media outlets, usually in fierce competition with each other. The readily available media coverage gave birth to new tactics in the repertoire of CLI terrorism: kidnappings, hijackings and hostage-takings. As one can easily imagine, without the attention of at least countrywide mass media, such actions would have come to the immediate attention of a very limited public only – which would have greatly impinged on their effectiveness. This is probably the reason why previous terrorist groups usually ignored them.

The new terrorist tactic of hostage taking swiftly led to yet another tactical innovation: the hijacking of airliners, a tactic also known as *skyjacking*. Taking their cue from a spate of criminally driven skyjackings in the USA during the 1950s/1960s, three terrorists of the Popular Front for the Liberation of Palestine (PFLP) hijacked El Al flight 426 on its way from Rome to Tel Aviv on 23 July 1968. The skyjackers forced the cockpit crew of the Boeing 707 to divert to Algiers, where a 40-day hostage crisis enfolded. In the end, the ten crewmembers and 38 passengers were released, and the hijackers were allowed to leave the country unhindered. Hijacking planes rapidly became quite popular, especially among left-wing terrorist groups: the plight of a comparatively large group of usually Western civilians crammed together in the confined space of an airliner for days was a guarantee for 24/7 media reporting – and thus for the terrorists a guarantee that their message found a global audience. It should be noted that airliners were and are not the only means of public transport falling prey to terrorist hostage-takers: ships, trains and buses have also been used for such a purpose, although only sporadically. The Dutch train hostage crisis of 2–14 December 1975, where seven persons of South-Moluccan origin kept 50 passengers as hostages for twelve days is a case in point.

Assassinations, kidnappings, and hostage-takings – these CLI tactics were and still are favoured by left-wing terrorist groups. But there are other and deadlier tactics, such as bombing attacks. Random bombing attacks targeting civilians can be traced back to the anarchist terrorists who were active before and after the end of the 1st World War. Italian-born anarchist terrorist Mario Buda for example detonated the first vehicle bomb (a horse-drawn wagon) on the corner of Wall Street and Broad Street in New York, 16 September 1920, presumably to avenge the famous Italian-born anarchists Nicola Sacco and Bartolomeo Vanzetti, who had been arrested a couple of months before. The detonation of *Buda's Wagon* (the term was coined by Davis 2007) resulted in the death of 38 civilians and the injury of approximately 140 others. Buda's improvised vehicle bomb was an act of indiscriminate violence, and as such already belongs to the category of conventional high impact (CHI) tactics. With regard to such CHI incidents, the availability of reliable explosives other than the dangerous (for the user) nitro-glycerine, trinitrotoluene (TNT) and dynamite resulted in a proliferation of bombing attacks after the end of the 2nd World War. Throwing crude bombs or 'Molotov cocktails' at a chosen target from a short distance quickly fell out of favour because of the risk involved for the operative as well. Instead, time-fused or, later, remotely-triggered land mine-style bombs – known as *improvised explosive devices* (IED) – came into fashion, while the car bomb took a little longer to establish itself as a weapon of choice: it first came to global attention with the devastating bombing of the US Marines headquarters in Beirut on 23 October 1983 (killing 241 soldiers), followed by the Oklahoma City bombing of 19 April 1995 (killing 168 people), and the first Bali bombings of 12 October 2002 (killing 202 people). Today, this kind of weapon is known as *vehicle-borne improvised explosive device* (VBIED). If prepared by expert bomb makers, both IEDs and VBIEDs can reach such a grade of sophistication and perfection that they are hard to detect even for experienced bomb disposal specialists (on the history of VBIEDs see Davis 2007).

The VBIED attack on the US Marines barracks is of importance for yet another reason: it was the first terrorist suicide bombing in history. The driver of the truck drove through the barbed-wire perimeter, slamming the vehicle into the lobby of the headquarters building, and then detonated the more than five tons of explosives. A little later, a second suicide bomber drove his truck bomb into the underground garage of a French barracks building and detonated it there. 58 French soldiers were killed in this terrorist attack. The attacks of 9/11, London 7/7 and the first Bali bombing of 12 October 2002 also were suicide bombings. Although tactically they were quite different (planes as VBIEDs in the case of 9/11, one car bomb and a rucksack bomb in the case of the Bali bombing, and four rucksack bombs in the case of London 7/7), these incidents clearly show how successful and devastating such operations can be. Apart from rucksack bombs or VBIEDs, another and more common type of suicide attack is the use of so-called suicide belts (also known as explosive belts) by a specially trained individual, male or female. The suicide belt is an invention of the Liberation Tiger of Tamil Eelam (LTTE) and was first used by a female LTTE Black Tiger – this was the name of their specialized, highly trained and highly motivated suicide unit – on 21 May 1991. On that day, Thenmuli Rajaratnam, nom de guerre Dhanu (aka Gayatri) killed former Indian Prime Minister Rajiv Gandhi who was fighting for re-election, at a rally in a small town near Chennai in Tamil Nadu/India using such a device (Pape 2005, 226–230). In this context, it should be pointed out that LTTE's Black Tigers carried out 76 of the 315 suicide attacks between 1983 and 2003, investigated by Pape between 1983 and 2003, "more suicide attacks than Hamas" (Pape 2005, 4; see also Pedahzur 2005, 71–86). The secular and vaguely Marxist LTTE firmly belonged to the camp of ethno-nationalist groups, driving home the point that not only religiously motivated groups are interested in indiscriminately killing people by way of suicide terrorism.

If the aim of the terrorists is killing as many people as possible in so-called mass-casualty attacks, it seems to be quite logical that they may indeed be interested in chemical, biological, radiological and nuclear (CBRN) attacks that would fall into the unconventional high impact or UHI category. It should be noted that with the exception of nuclear devices, the other three elements of the acronym CBRN have already been used, or at least threatened to be used. The Japanese messianic-apocalyptic cult Aum Shinrikyo for example used self-produced Sarin gas, a chemical agent, against the subway system of Tokyo on 20 March 1995, killing 12 people and seriously injuring a further 54, while hundred others were also affected (for more information on Aum Shinrikyo, see Brackett 1996). Biological agents have been used in the immediate aftermath of 9/11 in the shape of anthrax. On 18 September 2001 five letters containing military-grade anthrax spores were mailed from Trenton/New Jersey to the offices of the New York Post, ABC, CBS and NBC and on 9 October two more were sent from the same location to US senators Tom Daschle and Patrick Leahy. As a result of anthrax inhalation, five people died, while seventeen others survived the attack seriously injured. The perpetrator of the attack was never clearly identified, although the prime suspect for the attack was Dr. Bruce Ivins, previously employed at US Army Medical Research Institute of Infectious Diseases (USAMRIID), Fort Detrick, Maryland. However, on 29 July 2008 Dr. Ivins committed suicide before he could be prosecuted (FBI, undated; on biological agents in general see Kellman 2007)). As concerns radiological devices, also known as 'dirty bombs', they have not yet been used by terrorist organizations for an attack. But in November 1995, a Chechen organization demonstrated that it could make use of such a device by burying radiological material (Caesium-137) in Moscow's Izmaylovsky Park (IAEA 2002). In December 1998, a Chechen intelligence service even claimed to have discovered a radiological device near a major railway line. The evidence presented is unconvincing, however. Another not entirely convincing case in the West is the one of José Padilla aka Abdullah al-Muhajir who was arrested on 8 May 2002 because intelligence services believed that he was working on a dirty bomb to be detonated in a US city on behalf of Al Qaeda. At the time of his arrest, Padilla had not yet acquired any radiological material needed for the device. Thus, some experts doubt that he was actually planning such an attack (Mariner 2005).

These examples demonstrate that the possibility of a UHI/CBRN attack should by no means be discarded without careful study. As already mentioned, the case of the so-called *Alphabet Bomber* Muharem Kurbegovic for example demonstrates that even a 'lone wolf terrorist' may under certain conditions have the means to launch a chemical attack: on 15 August 1974, and after already having killed three people and injured another thirty-five in a conventional bombing attack at Los Angeles' LAX Airport on 6 August, Kurbegovic threatened to launch a Sarin gas attack targeting Washington's Capitol Hill on behalf of a (non-existent) group named *Aliens of America*: "Imagine what will happen if we are lucky and the wind blows from Supreme Court to White House to Pentagon" (as quoted in Simon 2001, 81). Fortunately, Kurbegovic could be arrested on 20 August before he could translate this threat into reality. In that case, 'we' were lucky while the terrorists were not. But as the IRA famously stated in the aftermath of the Brighton Hotel bombing of 12 October 1984, the terrorists need to be lucky only once, while counter-terrorism experts need to be lucky all the time. However, there are easier ways to carry out mass-casualty attacks – attacks that do not require untested and unreliable agents such as homemade Sarin gas or anthrax.

Probably the easiest way to kill as many people as possible while also inflicting serious psychological damage on the survivors, bystanders, citizens of the town or city affected, and attracting at least a nation-wide but potentially global audience, are gun assaults. Gun assaults are carried out by one (less frequent) or several (more frequent) persons equipped with small arms such as assault rifles and/or shotguns plus hand grenades or IEDs. This tactic that belongs to the CHI category was used by various terrorist organizations as early as 30 May 1972. Here, a three-person hit squad of the Japanese Red Army acting on behalf of the Popular Front for the Liberation of Palestine, External Operations (PFLP-EO) opened fire on disembarked passengers at Lod Airport (today Ben Gurion International Airport, Tel Aviv) in Israel, killing 24 persons and injuring a further 78. The rationale behind this attack was the knowledge that the Israeli security personnel only checked people of Arab origin – hence they failed to detect the assault rifles and hand grenades hidden in the suitcases of the three seemingly harmless Japanese travellers. Very similarly, the pro-Palestinian Abu Nidal Organisation (ANO) carried out a coordinated attack on the airports of Vienna and Rome on 27 December 1985, killing three people while injuring 44 in Vienna, and killing 13 people and injuring a further 76 in Rome (Seale 1992, 238/243–247). Gun assaults are also one of the tactics of choice of the Jihadist organization *Lashkar-e Tayyiba* (Army of the Pure), active mostly in Kashmir: between 1998 and 2003, but largely unreported by global media, terrorists of this group conducted several gun assaults, killing scores of people. The deadliest one however was the Mumbai attack on 26–29 November 2008, an incident that brought global media coverage, partially live. And this audacious CHI incident finally brings the narrative to the here and now.

2.2 Cities as Targets: Mumbai, Paris and Brussels

The multiple co-ordinated and complex 'marauding attacks' (as they are now called) of Mumbai, Paris and Brussels indicate that arguably the most serious terrorist threat in the present and for the foreseeable future (notwithstanding my exhortation to keep our minds open) is the one posed by small cells of terrorists belonging to, or being loosely affiliated with, either Al Qaeda or ISIS and striking on their behalf. In the case of the three-day Mumbai siege that cost 166 individuals their lives while injuring more than 600 others, the ten perpetrators belonged to the just mentioned terrorist group Lashkar-e Tayyiba. It later emerged that in a case of what Kilcullen (2016, 119) calls 'expeditionary terrorism,' they infiltrated by sea from Pakistan, capturing an Indian trawler and murdering its crew before making landfall in Mumbai's Sassoon docks. The terrorists and their Pakistan-based handlers had selected their targets with great care: the upmarket Oberoi Trident and Taj Mahal Palace & Tower hotels plus the Leopold Café were chosen because these are places where foreigners and middle-class Indians could be found (and killed), while the Nariman House was chosen because it housed a Jewish outreach centre. The Chhatrapati Shivaji Terminus as Mumbai's main railway station and the Cama Hospital also made it on the final list of targets since these places were perfect to maximize casualties with only low risk to the perpetrators.

Police investigations in the aftermath of the event revealed that the ten members of the strike team had been extremely well prepared for sustained action: prior to the attack, they had reconnoitred their targets and the streets connecting them to gain a clear understanding of entry and exit route, and of the number and capabilities of security forces deployed in the vicinity. They had trained for and rehearsed the attack several times, under the guidance of experienced instructors. On the eve of the attack, they equipped themselves with plenty spare ammunition for their Kalashnikov-type assault rifles, while also packing hand grenades and explosives. Furthermore, they thoughtfully included water, food, and drugs to stay awake and active as long as possible. While carrying out their attacks, they observed strict fire discipline, firing in short, controlled bursts instead of just emptying their magazines, as untrained actors probably would have done. Throughout the event, they also stayed in close touch with their handlers via mobile phones, thus always being up-to-date about impending counter measures of security forces, or of the whereabouts of numbers of civilians hidden away in various locations in the two hotels. This enabled them to continuously out-manoeuvre police units trying the close in on them, while capturing and killing individuals who otherwise would have escaped. All these sensitive and extremely valuable information was provided by live TV reporting and numerous live radio interviews by TV and radio stations opting for sensationalist coverage of events instead of a responsible one. For example, when Special Forces abseiled from a helicopter onto the roof of Nariman House, this was broadcast live on TV – thus allowing the terrorists to prepare for the impending attack (discussions and evaluations of this incident are offered for example by

Acharya et al. 2009; Azad and Gupta 2011; Riedel 2012; Scott-Clark and Levy 2013).

The combined CHI tactics used in this event were very similar (albeit much larger in scale) to the so-called *Fidayeen* (lit.: 'those who sacrifice themselves') attacks in Kashmir, which has a long history of drawn-out gun battles between highly mobile two-men teams equipped with assault rifles and hand grenades on the one side, and Indian military or paramilitary units on the other. Apart from an unsuccessful attack on the Indian Parliament (Lok Sabha) in December 2001, this was the first time that this tactic was employed outside of the Indian federal state of Jammu and Kashmir. From a terrorist's perspective, the Mumbai shootings constituted CHI terror at its best: lots of people were watching, while lots of people were dying. And compared to 'high end' terrorism such as 9/11, Mumbai-style marauding attacks consisting of mass shootings combined with bombings are much easier to plan, to train for, and to carry out. Consequently, many terrorist experts feared that this advanced CHI tactic would swiftly be copied by other terrorist organisations as well, and used to attack our own cities, and our own vulnerable open societies. Nevertheless, it took no less than seven years until another terrorist outfit could launch an attack against a major Western city: on 13/14 November 2015, nine terrorists affiliated to ISIS (which in the meantime had overtaken Al Qaeda as the leader of the Islamist cause) successfully attacked multiple locations in the city of Paris. Like the perpetrators of the Mumbai attacks, the terrorists opted to attack several carefully selected venues with a combination of shootings, bombings, and a hostage crisis to maximize both casualties and media exposure. The intention behind these attacks was to take revenge for the French air strikes on ISIS positions in the Middle East. After the attack, ISIS made it known that they purposely selected venues where ordinary people went on a Friday night to enjoy the weekend: 'destroying their beds, making their lives bitter' allegedly was the motto of this incident. Hence, the locations chosen for the multiple strikes included the Stade de France that was filled with 80,000 spectators (amongst them President Hollande) watching a friendly match between France and Germany, several bars and restaurants, and the Bataclan theatre (on the incident, see for example Chauzal et al. 2015; ICT Herzliya 2015).

The incident began at 9:20 pm at the Stade de France with a suicide bombing, followed by two others in a matter of minutes. It later emerged that the three suicide bombers had hoped to get access to the stadium to explode their suicide vests on the main stands, or alternatively detonate them outside amid a crowd trying to get in. Fortunately, the security measures in place prevented them from entering: the first suicide bomber who tried was patted down and denied entry when the security guards noticed a suspicious device – his suicide vest. He then detonated his vest at the gate but managed to kill only one person apart from himself. The other two suicide bombers failed to kill anybody but themselves, making this first incident a failure. Sources from the French police speculated that the timing of this 'badly organized' suicide bombing was wrong: when the suicide bombers appeared twenty minutes into the match, the spectators were already inside the stadium, and the squares in front of the gates were empty. The match itself continued, with most of the spectators unaware of what happened until the end, although two of the three detonations were clearly audible. The French Minister for Sports, Patrick Kanner, later claimed that the match was allowed to continue in order to prevent a panic. In any case, it is strange that the suicide bombers did not abort their mission in favour of attacking other suitable targets.

The remaining six members of the strike team were far more successful. In a drive-by shooting type of attack that commenced at 9:25 pm and ended eleven minutes later at 9:36 pm, two of them systematically fired first at staff and patrons of the Cambodian restaurant *Le Petite Cambodge* and the bar *Le Carillon* at the Rue Alibert/Rue Bichat intersection (9:25 pm), then at the staff and patrons of the café *A la Bonne Biere* and the pizzeria *Casa Nostra* (9:32 pm), and finally at the guests and staff of the restaurant *La Belle Equipe* (9:36 pm). A suicide bomber attacked the Restaurant *Comptoir Voltaire* at 9:40 pm. All taken together, 39 people were killed in these incidents, while a further 30 were severely injured. At the same time the *Comptoir Voltaire* was bombed, another three-men team of the ISIS-affiliated terrorists arrived at the Bataclan theatre which was filled with about 1500 people, listening to the band *Eagles of Death Metal.* Just like the Lashkar-e Tayyiba team attacking the main railway station in Mumbai seven years before, they observed strict fire discipline, calmly firing in short bursts, reloading their magazines, then firing again. According to eyewitnesses, they also lobbed hand grenades into the crowd during their twenty minutes shooting spree. When the first police patrols arrived at 10:00 pm, the terrorists barricaded themselves in the theatre with a substantial number of hostages – depending on the sources, between 60 and 100. Special police forces finally stormed the venue at 00:20 am in a well-planned operation that lasted just three minutes. All three terrorists were killed in the counterattack that probably saved many hostages their lives. Nevertheless, 89 people were killed during the Bataclan siege, and several hundreds were injured. As a further result of the Paris attacks, many public venues inside and outside of France were closed for security reasons, major sports events were cancelled, the French borders were temporarily closed, emergency law was declared, and Article 47.2 of the EU Treaty of Lisbon – the article that deals with common defence – was invoked. Kilcullen (2016, 220) astutely sums up the significance of the events as follows:

> The Paris attacks were, in some ways, more concerning than 9/11, though casualties were much lower, because rather than a one-off attack from outside the country (the 'expeditionary terrorism' model), they signalled the existence of a paramilitary underground – a better-organized, more capable version of the ISIS Internationale though not yet at the level of a full-blown *wilayat* [province] – operating in cities within France, Belgium and Germany, and possibly also in Denmark and the Netherlands. Early indications suggested that Paris might represent the start of a sustained urban guerrilla campaign – and perhaps not just in Europe.

The next Mumbai-style combined CHI attack against a Western city followed about four months later, this time targeting Brussels, the capital of Belgium and the seat of the European Union. This incident was carried out by five ISIS supporters linked to the Paris attackers. They targeted the airport and one subway station (Maalbek metro station) in an operation lasting one hour 13 minutes. Three suicide bombers attacked the first target, Brussels International Airport, at 07:58 am.

Although one of the three seemed to have lost his nerves and aborted the attack, the other two detonated their explosives hidden in suitcases as planned after firing at passengers at check-in with AKM-type assault rifles. The other two team members detonated their suicide vests at 09:11 am at the Maalbek metro station. All in all, 32 people were killed in these two incidents, and about 316 were injured. The rationale behind the target selection apparently was to shut down the city's transport systems and to kill as many people as possible. However, it later emerged that the Brussels-based perpetrators were preparing yet another attack on Paris, and only chose to switch to these two hastily selected targets in Brussels on short notice because they feared that the police was closing in on them, and that they were running out of time. Like in the case of Paris, many public venues were closed in the aftermath of the attack, and several sports events were cancelled or at least postponed – thus driving home the point to any interested terrorist group just how effective such attacks are with respect to paralyzing everyday life, and at least temporarily shutting down substantial parts of our open societies.

Against the backdrop of these incidents, it thus can be argued that, just as Homer-Dixon predicted, terrorists have mastered the art of jujutsu-like 'complex terrorism' by now, and that they now have the capability to "bring modern, high-tech societies to their knees" (Homer-Dixon 2009). John Robb makes a very similar argument, stating that terrorists have developed an "ability to fight nation-states strategically – without weapons of mass destruction." He calls this *systems disruption,* described as "a simple way of attacking the critical networks (electricity, oil, gas, water, communications, and transportation) that underpin modern life" (both quotes from Robb 2007, 5). Although he made his arguments against the backdrop of the insurgency in Iraq, the events of Mumbai, Paris and Brussels indicate that at least a partial and temporary systems disruption in the shape of "infrastructural terrorism" (Graham 2010, xxiv) is in the reach of terrorist groups. And with that remark, it is time now to take a closer look at this now manifest interest of terrorists in our cities from a theoretical angle.

2.3 Urban Guerrillas: Marighella Revisited

Post-9/11, many books and articles have been published addressing basically all possible aspects of terrorism and counter-terrorism, but in my opinion, the following three books read together probably best illuminate the most probable future of urban terrorism as well as counter-terrorism: Kilcullen's *Out of the Mountains* (Kilcullen 2015), Graham's *Cities Under Siege* (Graham 2010), and Marighella's *Minimanual of the Urban Guerrilla* (Marighella 1969). To start with the latter, and in the light of Mumbai, Paris and Brussels, one could well argue that Marighella was ahead of his time when he wrote the minimanual. Today, his recommendations make much more sense than in the early 1970s when most left-wing guerrillas and terrorists still were deeply influenced by Maoist doctrines, according to which cities should initially be ignored and then increasingly be isolated from each other via the already conquered rural environment, now dominated by the communist proletarian peasant masses (Burton 1975, 73). Even Régis Debray's famous *foco* theory, which revolves around a small guerrilla group establishing a foothold in a remote rural area as the focal point of the revolution, was more of a reinterpretation of contemporary mainstream ideas about guerrillas and their tactics than a completely new approach (Burton 1975, 76–77). In Debray's opinion, a premature "descent to the City" was not just a waste of time and lives but would end in "uncontrolled terrorism in the city and banditry in the countryside" (Debray 1967, 75). Residues of the original Maoist doctrine admittedly are also evident in Marighella's otherwise novel thinking: for him, urban guerrillas basically were auxiliaries of the rural guerrillas, and subordinated to them. Nevertheless, unlike Debray and others, he acknowledged the fact that in his time, a move from rural areas to the cities had already begun – a move that steadily decreased the population living on the countryside, thus gradually eroding the pool of potential supporters for the rural guerrillas. Marighella was aware of this process, and of the defeat, capture and execution of Ernesto 'Che' Guevara in Bolivia on 8/9 October 1967 by US-trained and equipped Bolivian Rangers. Hence, his view is more nuanced, trying to reconcile old and new:

> The urban guerrilla is a person who fights the military dictatorship with weapons, using unconventional methods. […] The area in which the urban guerrilla operates is in the large Brazilian cities. […] The primary task of the urban guerrilla is to distract, to wear down, to demoralize the military regime and its repressive forces, and also to attack and destroy the wealth and property of the foreign managers and the Brazilian upper class. [… His] aim is to aid the rural guerrillas and to help in the creation of a totally new and revolutionary social and political structure, with the armed population in power. (Marighella 1969)

This implies that for Marighella, the city was only a small part of a much wider battlespace that extended to larger rural areas, to the whole country, and, finally, to the whole continent. Marighella's definition of the urban guerrillas (as well as his whole minimanual) was eagerly read and adapted to many different environments – including environments where rural guerrillas were notably absent. One example from the European scene would be the 'Fighting Communist Organizations' (FCOs) of Western Europe in the 1970s such as the Italian *Brigate Rosse,* the German *Red Army Faction* (RAF) or the French *Action Directe* – all of them organizations solely fighting in the cities as this was the place where both their enemies (representatives of the political, economic and judicial systems as facets of the state) and their potential supporters (citizens meant to be revolutionized) were to be found (Taw and Hoffman 1994, 7–8). What hampered the smooth translation of Marighella's suggestions which include weapons, tactics, fire teams and the like into tangible action was the fact that when his minimanual was published in 1969, urbanization was in its infancy, especially in the West where 'megacities' of more than eight million inhabitants were few and far between, and where 'hypercities' with over 20 million inhabitants (Davis 2006b, 5), left alone megalopolises defined as a clustered network of cities with more than 25 million citizens (Gottmann 1989, 163), still were notably absent. Even more importantly, 'no go' areas out of reach and control of everyday policing did not yet exist, nor was there a rural guerrilla which could have provided material and moral support. For FCOs such as the RAF, it was thus rather difficult to operate in a city undetected long enough to inflict such damage as mentioned by Marighella – all they could usually achieve was one single operation before they had to move on in order to escape the police's incessant dragnet search. For this reason, many scholars actually disputed the existence of 'urban guerrillas' as a contradiction in point. For example, in 1974, a sceptical Håkon Wiberg stated that "the best proof for the possibility of a successful urban guerrilla would be a concrete case; and this has not yet been demonstrated" (Wiberg 1974, 19), while Walter Laqueur dismissed the validity of the concept out of hand:

> 'Urban guerrilla' – a term as widespread as it is fraudulent – is indeed urban, but it is not 'guerrilla' in any meaningful sense of the term. The difference between a guerrilla and a terrorist is not one of semantics but one of quality. A guerrilla leader – to put it in the briefest possible way – aims at building up ever-growing military units and eventually an army, and establishing liberated zones in which an alternative government can be put up and openly conducted. All this is quite impossible in cities; the different environment dictates a different strategy and this has basic political consequences. While guerrillas have quite frequently used terrorist tactics [...], the opposite has virtually never happened, for in the urban milieu there are no opportunities for guerrilla warfare. (Laqueur 1987, 5)

Laqueur was quite right: even in the late 1980s, urban guerrilla warfare as theorized by Marighella did not yet exist. What did exist were either spontaneous riots such as the race riot of Newark, New Jersey on 12–17 July 1967 (26 people dead, several hundreds injured) on the one hand, or larger-scale insurgencies such as civil wars that also swapped, at least occasionally, into towns and cities on the other. An example for the latter would be the Second Vietnam War, especially the Tet Offensive (30 January – 23 September 1968) and the battles of Hue (30 January – 3 March 1968) and Saigon (30 January – 7 March 1968). For such occasions, Burton explains the mechanics of urban operations:

> If the urban insurgent was [...] to harass the troops sent against him he would have to utilize the mobility which knowledge of the terrain (and ideally the support of the people) gave him in order to move both laterally and vertically through the buildings, firing down on the troops who would, in such circumstances, find it difficult to bring their superior firepower to bear. Barricades would be used, not for shelter but to tempt the troops sent to dismantle them into a killing zone overlooked by small trained bands of insurgents. (Burton 1975, 3)

Until rather recently, probably definable as Mumbai 2008 as the benchmark event in that regard, conditions conducive to such operations rarely existed, and most definitely not in a Western environment. It is true that Al Qaeda-run training bases in Afghanistan of the late 1980s, such as the Khaldan camp, already included urban warfare as well as urban sabotage attacks against "electric plants, gas plants, airports, railroads, large corporations, gas, gas installations and military installations also [and] ... hotels, where conferences are held" (Ressam 2002, 173), but these courses were geared towards Afghan-Arabs about to return to their own home countries, for example Libya and Algeria, in order to continue their jihad at home, but not (yet) intended for those interested in waging urban guerrilla warfare in Western countries. Interestingly however, the German experience shows that even the comparatively few terrorist attacks that did occur during the 1970s and 1980s proved to be enough to create an atmosphere of intense and all-pervasive fear. One of Germany's foremost authors, Günter Grass, even sarcastically spoke of a 'war of six against sixty million', while Germany's tabloids such as the *Bildzeitung* habitually and without any trace of sarcasm used the 'war' paradigm to make (conservative) sense out of what was happening. With the wisdom of hindsight, one could see the shadow of the future in the New Left terrorists' actions – even though their actions were of a CLI nature that avoided death and injuries among innocent bystanders, and thus rather harmless when compared to the mass casualty attacks of Al Qaeda and ISIS.

Of course, FCOs were not the only groups closely following Marighella's suggestions. In a British context, the IRA's Bishopsgate (24 April 1993) and Canary Warf (9 February 1996) bombings need to be mentioned, and the audacious mortar bomb attacks on 10 Downing Street on 7 February 1991 and London Heathrow International Airport between 9 and 13 March 1994 as well. Again, in relation to these actions and counter-actions such as the famous 'Wall of Steel' followed by the 'Wall of Glass' (which will be discussed in some detail in the next part), one could see the shadow of the future. And from a scholarly perspective, some farsighted experts took note of the rising rate of urbanization in the West that would, in the future, provide the necessary background for urban guerrillas postulated by Marighella. Writing in 1974, Burton for example maintained that "[there] is no doubt however that economic and demographic changes are producing an environment increasingly sensitive to the use of terrorist techniques" (Burton 1975, 239–240). The events of Paris 2015 and Brussels 2016 indicate that Burton's future has finally arrived, and that the proper environment conducive to urban guerrilla warfare now seems to be present.

2.4 Beyond Marighella: Modern Urban Guerrilla Warfare

The future of urban guerrilla warfare (our present) arrived in the shape of multiple coordinated mass-casualty attacks launched by a new wave of terrorists – a new wave of religious terrorism with the two Salafist-Jihadist movements of Al Qaeda and ISIS at the centre of it. Contrary to earlier manifestations of terrorism, this new variant explicitly aims at causing as many deaths and injuries as possible in multiple, coordinated mass-casualty attacks targeting Western cities, Western cosmopolitanism, the Western way of life, and even individual pursuits of happiness – aiming at 'destroying [our] beds, making [our] lives bitter,' as ISIS explicitly stated. Apart from killing or injuring as many people as possible, they also aim at bringing Western daily lives to a stand-still by forcing authorities to temporarily shut down air, land, and (depending on the location) sea transport systems, and to cancel or at least postpone public events for security reasons. Brussels International Airport as the sole airport of this capital for example was shut down for no less than 12 days in the aftermath of the attack, while public events such as Christmas markets were cancelled.

Of course, it goes without saying that if a terrorist group or movement wants to conduct mass-casualty attacks that will be covered live on TV, radio and social media such as Facebook, Twitter or Periscope, they need to go where the people are. This argument alone makes cities eminently vulnerable – busy shopping streets and underground stations/train stations during rush hour readily come to mind, and so do the London 7/7 suicide bombings or the 11 March 2004 Madrid train bombings. But selecting cities as their main targets usually goes much further than that. For fundamentalist Islamists, such as Al Qaeda and ISIS, or for any fundamentalists whatever their religion, cities are intrinsically sinful places, cursed by God. In Al Qaeda's rhetoric, cities are "unparalleled concentrations of sin, debauchery, greed, materialism and soullessness" (Graham 2010, 41). Hence, for religious reasons as well, cities are perfect targets for them. The same can be said about tourist resorts and destinations such as Luxor/Egypt (shooting attack 17 November 1997, 62 people killed, suicide bombing attack foiled on 10 June 2015), Kuta Beach/Bali (suicide bombing attack 12 October 2002, 202 people killed, attacked again 1 October 2005, 20 people killed), Sousse/Tunisia (shooting attack 18 March 2015, 39 killed), or the attack on the gay nightclub in Orlando/Florida for that matter (shooting attack 12 June 2016, 49 people killed) since these also are locations where Western people unashamedly follow their hedonistic lifestyle that, for Islamists, smacks of *jahiliyyah*, i.e. the pre-Islamic age of ignorance.

Apart from being convenient targets, modern major cities as 'urban jungles' offer yet another opportunity for terrorist groups intent on attacking them: unlike their smaller avatars of the 1970s and 1980s, at least some of them now offer safe havens in the shape of economically and socially deprived suburbs and/or inner-city districts (depending on the city in question) out of day-to-day control of the police, where terrorist cells can plan and prepare their attacks, and carry them out. In these areas, they also find ready support networks of sympathizers that can provide the terrorists with forged ID cards and passports, vehicles with forged number plates – and possibly even with sophisticated weaponry including state-of-the-art improvised explosive devices (IEDs) including suicide belts or vests. And if the terrorists' chosen tactic does not involve suicide bombing, the urban jungle also offers them the opportunity to disappear, seemingly without trace: since the times of the FCOs, most Western European cities have witnessed a steady growth, while also getting more heterogeneous due to the influx of high numbers of migrants from within and without Europe. Driven by cultural, religious and linguistic differences, the bulk of these various immigrant groups chose to live in close proximity to each other, thus resulting in a gradual ghettoization. Well-known due to the events in Paris 2015 and Brussels 2016 are the already mentioned suburbs (banlieues) of many French cities, especially the belt surrounding Paris, and the inner-city district of Molenbeek in Brussels, adjacent to the Gare du Midi and thus geographically in the heart of this capital but effectively isolated from it. Similar ghettos can be found in most other Western European cities as well. In the light of the current mass migration of people from the Middle East and elsewhere fleeing from terrorism, authoritarianism and/or civil wars – a mass migration that brought more than one million refugees to

Germany alone – it can be taken for granted that these ghettos are bound to grow, and the problems they bring with them as well.

These problems are manifold, but a high rate of unemployment, especially youth unemployment, can be identified as one of the main issues (on this issue, see for example Bay and Blekesaune 2002). Another one is the reluctance to embark on meaningful integration shown by a high percentage of immigrants who are either not willing or not able to adjust to what they see to be an alien culture, but also by a certain percentage of locals who tend to look down on the immigrants and to discriminate and marginalize them – which is one of the reasons behind the high rate of unemployment. For second and third-generation immigrants who find themselves betwixt and between different cultures and lifestyles, the situation they find themselves in is oftentimes perceived as hopeless, with no way out. It is thus not surprising that some of them are susceptible to anything that promises them an alternative. This alternative can be becoming a gang member, a drug dealer, or a so-called 'hit-iste' as someone leaning against a wall during most of the day ('hit' = Algerian Arab for 'wall') – which refers to an urban (male) poor settling into a life on the dole. Listening to the messages of Islamist preachers of hate and subscribing to their cause is just another alternative. Some of them turn into what sociologist Farhad Khosrokhavar (2009) called 'negative heroes': disoriented and frustrated young males who kill to just once feel a sense of triumph and greatness. All of them are prone to express their frustration by way of political violence below the threshold of terrorism, given the right trigger. One such example are the series of riots starting in one of Paris' suburbs, Clichy-sous-Bois, on 27 October 2005, quickly spreading to suburbs of Lille, Lyon, Marseille, Strasbourg and Toulouse. The trigger for these events was the death of two French youths of North African descent, Bouna Traore (15 years old) and Zyed Benna (17 years old) who died of electrocution when they fled into a substation to hide from the police. The series of riots that included general violence, looting and arson lasted until 16 November 2005. 126 police officers and firefighters were injured, two civilians were killed by the rioters, and 2.888 people were arrested (see Cesari 2005 for further information). The second example are the London Riots of 6–11 August 2011. The trigger event of these riots that also included general violence, arson and looting was the fatal shooting of Tottenham resident and suspected drug dealer Mark Duggan by the police on 4 August. The riots started two days later when an initially peaceful protest march went out of control.

It should be noted that only comparatively few of these disaffected youths take part in such events, while even fewer cross the line from non-lethal, spontaneous political violence as such to lethal and carefully planned acts of terrorism. Nevertheless, as Molenbeek shows, the groundswell of frustration and anger towards 'the system' or 'the man' combined with a tradition neither to trust nor to speak with the police allows those who do cross the line and become either criminals or terrorists to freely move in these areas of the urban jungle – areas which are no longer under the firm control of everyday policing. Affected police forces routinely deny this, of course, but this denial does not change the fact that individual police patrols now have reason to fear for their lives if they try to stop a car for traffic violations in the wrong part of the city – a part for example dominated by extended families or clans who can, if need be, muster dozens of male family members at the scene after some phone calls to face down police officers not necessarily prepared for such encounters. Indicative for that is an incident that occurred in a part of the German city of Gelsenkirchen (Ruhr area) in 2015. Although the local police understandably were reluctant to call this part a 'no go' area controlled by a Lebanese clan, they did admit that they only went there in force, supported by Bereitschaftspolizei (paramilitary riot police) units – while members of the Lebanese clan allegedly claimed that they would win a stand-off because there were too many of them, and because the federal state of North Rhine-Westphalia did not have the finances to deploy as many police as required to win this confrontation (Biernat 2015). Quite ironically, while upper and middle-class segments of citizens have their own gated communities under the control of, and protected by, private security providers as we shall see, some criminal sections of society also seem to have taken over parts of the cities, controlling and 'protecting' them with their own force of thugs. As German journalist Christoph Wöhrle explains it, "they live in a world created by themselves, to which police, authorities or politics have no longer access" (Wöhrle 2017, translation mine). From an academic perspective, and according to Fall (1964) and Kilcullen (2015), this phenomenon can be described as a 'system of competitive control' between different sets of stakeholders.

It is however not only these lesser-accessible parts (for law enforcement) of the urban jungle that allow the modern urban guerrilla to effectively drop off the radar screen. Rather, the hustle and bustle of modern sprawling cities in general renders it difficult for counter-terrorism forces to keep track of their movements – even if they manage to detect them. As the US Defense Intelligence Reference Document (DIRC) pointed out in a somewhat wider context:

> Opposition forces will camouflage themselves in the background noise of the urban environment. Within the urban environment, it is not the weapon itself but rather the city which maximises or mutes an arm's effectiveness. In claustrophobic alleys and urban canyons, civilians are impossible to control or characterise as friendly or not. Weapons hidden beneath a cloak, in a child's carriage, or rolled in a carpet, can get past security undetected. (as quoted in Graham 2010:159)

In itself, this is nothing new and precedes the advent of Islamist terrorism by a couple of decades. For example, during the famous 'Battle of the Kasbah,' the Algerian FNL used females to carry explosives, concealed beneath their clothes, through very similar 'claustrophobic alleys' either to the FNL operators designated to carry out an attack, or to the location of the attack itself. The vast scale of modern sprawling cities of today however makes it arguably much easier to disappear in the anonymous masses, and to retreat into already existing 'no go' areas. Hence, they now seem to be much closer to fulfil the criteria of being 'guerrillas' in the academic sense of the word. One might of course still argue with Laqueur (1987, 5) that these 'no go' areas are of a temporary nature, and that this "equilibrium" will not last long – but the same can be said about the liberated areas under the control of 'real', i.e. 'rural' guerrillas.

Somewhat ironically, there might be a price to pay for terrorists as well: coordinating attacks are not necessarily made easier with hordes of pedestrians or endless queues of cars trapped in equally endless traffic jams inhibiting rapid trips from one part of the city to another. A fitting example for that is the terrorist attack on 17 August 2015 in Bangkok targeting the popular *Erawan Shrine* at the Ratchaprasong Intersection (in close proximity to several 5-star hotels and up-market shopping malls) and the new *Asiatique* Night Bazar at the Chao Phraya river front. The incident did not involve suicide bombings – rather, the bombs hidden in backpacks were meant to be left behind at the chosen locations, with the pre-set timers allowing both terrorists to escape before the detonation. However, the operator tasked with carrying out the Asiatique attack apparently got stuck in the early evening rush hour traffic and had not yet reached his destination when the Erawan Shrine bomb detonated. Hence, faced with the prospect of ramped-up security on the site, he aborted the attack, ditching the bomb near a busy pier. This example of Murphy's Law notwithstanding, on balance it is probably fair to say that modern cities – especially those 'megacities' with a population surpassing 25 million – as veritable 'urban jungles' offer an intricate mix of opportunities and challenges which, taken together, can foment violent protest, including terrorism.

In the case of European cities, it is worthwhile to repeat some numbers pertaining to ISIS here: as the International Centre for Counter-Terrorism at The Hague reported with regard to foreign ISIS fighters coming from the European Union, between 90–100 percent of about 4000 (3922–4294) of them hail from urban areas – and about 30 percent or about 1200 already returned home (Mehra 2016, 3; also see Reed, Pohl and Jegerings 2017). Again, this does not mean that all of them are battle-hardened, hard-nosed fighters willing to carry on their jihad on their home turf: some of the returnees are traumatised by what they have seen, or what they have done, and many more are utterly disillusioned by all the savagery they witnessed. This still leaves a substantial number of potential terrorists – and their ranks are swelled by those who infiltrate into Europe using the concealment offered by the waves of immigrants fleeing from the turmoil in the Near East and other places on the one hand, and a certain number of self-radicalized sympathizers willing to move on from verbal support to tangible, terrorist action, just as the perpetrators of the 14 July 2016 Nice Massacre and the 22 May 2017 Manchester suicide bombing, Mohamed Lahouaiej Bouhlel and Salman Abedi, did. This, then, begs the question of how to respond to the threat posed by the new urban guerrillas (including 'lone wolves') ranging from the 'terrorism of opportunity' (as French President Hollande called it after the Nice attack) that uses everyday items such as knifes, axes, cars or trucks to complex-co-ordinated multiple strikes based on IEDs and modern weapons on the one hand, and to the new 'high end' terrorist urbanism in the shape of infrastructural terrorism on the other?

Chapter 3
Reactions: The New (Para-) Military Urbanism

Abstract In this chapter I argue that the new wave of urban guerrillas needs to be countered by appropriate counter-measures by law enforcement, paramilitary and military forces who also have to move into cities in order to defend them. I start the chapter with a more detailed introduction into the art of Critical Infrastructure Protection (CIP), commenting on the difficulties of protecting our open societies against modern forms of terrorism or 'urban guerrilla warfare'. Next, I critically discuss the two basic (ideal type) models on how to respond to the terrorist threat: the Criminal Justice Model (CJM) and the War Model (WM). Drawing on riots in a number of French banlieues (suburbs) and on the London 2011 riots, I also point out that terrorists are only the tip of the iceberg when it comes to the outbreak of latent conflicts in modern urban sprawls. I conclude this section with a first brief look at the question of how far liberal democracies can go in order to respond to the terrorist threat.

Keywords Military urbanism · Critical Infrastructure Protection (CIP) · Risk management, · Criminal justice model · War model · Liberal democracies · Law and terrorism

It is not only the urban guerrillas who have rediscovered the cities as a suitable battlespace, but also the military as well as paramilitary police forces – the latter mainly in response to modern terrorism. The militarization of urban space, or the "new urban militarism" as Graham (2010) calls it, is in itself nothing new as well and not necessarily a reaction to urban terrorism or urban guerrillas – or revolutionaries, for that matter. After all, as Burton observes, "Men have been fighting in the streets ever since the latter were built" – although he admits that "it was not, however, until the nineteenth century that physical and political conditions coincided to produce a theory of urban insurrection still relevant today" (Burton 1975, 2). Or as Taw and Hoffman (1994, 7) note, "[whether] as a catalyst or triumphant endpoint, the city has historically been the cynosure of all revolution. In the industrialized West [...] the city took on an all-encompassing role as a birthplace of, crucible for, and ultimate prize awarded to revolutionaries." Hence, securing the city always tended to be high on the agenda of the governments of the day. One could for example go back

P. Lehr, *Counter-Terrorism Technologies*, Advanced Sciences and Technologies for Security Applications, https://doi.org/10.1007/978-3-319-90924-0_3

to the reconstruction of Paris by Georges-Eugène Baron Haussmann from 1853 onwards, featuring amongst others wide boulevards that facilitated swift troop movement and that could easily be blocked by one cavalry squadron, and were dominated by roundabouts and squares as ideal emplacements for cannons, just in case the masses got restless again. He himself admitted that his renovation of Paris did not only aim at making the city more pleasant for its inhabitants, but also at "gutting old Paris, of the quarter of riots and barricades" ("l'éventrement du vieux Paris, du quartier des émeutes, des barricades", as quoted in Moncan 2012, 34). Baron Haussmann would have readily agreed to DIRC's fear of the 'claustrophobic alley'.

Ironically, shortly after Haussmann was dismissed by Emperor Napoleon III in early 1870, the Franco-German War triggered exactly such an event in Paris that he had tried to forestall: the uprising of the Paris Commune in 1871. It is however uncertain whether Haussmann's boulevards played a central role in the swift suppression of the Communards lead by Auguste Blanqui in just one weeks: it is much safer to argue that it was the fact that they were divided in several fractions, outnumbered five to one, without adequate weapons and without any chance to draw on support from the regions outside of Paris (Rougerie 2014, 115-117). Nevertheless, Burton is quite right to point out the following:

> After the defeat of the Paris Commune, the physical planning of cities and the increasing gap between the resources available to the citizens and those of the professional soldier rendered street warfare on Blanquist lines a quite hopeless, indeed suicidal, technique. Narrow, medieval streets of tall, huddled buildings gradually gave way to wide, straight prospects and cobblestones to tarmacadam. While the removal of paving stones deprived insurrectionaries of their building materials, the regular patterns of the new streets gave an excellent field of fire to the weapons placed in the hands of the soldier [...]. The advent of more accurate repeater rifles and, in particular, the machinegun rendered the city population a helpless mass quickly dispersed or, if need be, slaughtered. (Burton 1975, 3)

Of course, long before Haussmann's boulevards and squares, citadels with high curtain walls and well-protected gates situated at dominating locations within cities fulfilled a very similar function: protecting the rulers not only from the enemy without (i.e. outside the city walls) but also, and more often, from the enemy within (i.e. its own citizens). After all, urban elites frequently saw themselves threatened by the underclasses, whose numbers rapidly increased with the onset of the industrial revolution. Squalid living conditions and blatant social disparities ensured that there also was a steadily rising number of dissenters of various convictions and of various degrees of violent inclinations. As the frequent clashes in a number of French suburbs or the London Riots of 2011 demonstrate, this has not changed much over the decades. Hence, the US Marine Corps convincingly argued in 1998 that "cities historically are places where radical ideas ferment, dissenters find allies and discontented groups find media attention" (as quoted in Golan 2006). This also explains why this 'military urbanism' is not necessarily a new thing altogether – or at least not completely new. I will return to this issue later in this chapter. First, it is important to take a look at the theoretical backdrop to protecting our cities and our open societies, and then at some lessons learned in the context of urban guerrilla or urban terrorist campaigns.

3.1 Cities Under Threat: Protecting our Open Societies

Seen from a theoretical perspective, counter-terrorism essentially is an exercise in risk management – which is why it is necessary to take a look at that issue now. As already mentioned, API/NPRA (2004, 3) defines risk as "a function of consequences of a successful attack against an asset; and the likelihood of a successful attack against an asset." The second part of this definition, especially the terms 'consequences' and 'likelihood', already hints at the importance of conducting a proper risk assessment based on a criticality assessment, a threat assessment, and a vulnerability assessment. A very simple risk model illustrates the close relationship between criticality, threat, and vulnerability, and it also alerts us that these three dimensions of a proper risk assessment are not identical. For example,

a) parts of our infrastructure may be critical as well as vulnerable, but there is no manifest or plausible threat against them;
b) parts of our infrastructure may indeed be vulnerable and under a manifest or plausible threat, but are not deemed to be critical;
c) parts of our infrastructure may be critical and under a manifest or plausible threat, but not deemed to be vulnerable.

A risk exists only when and where these three variables overlap, and detecting this overlap is what a proper risk management is all about. On the basis of the debate of the previous chapter, it can be argued that there is, indeed, a clear and present danger to our cities posed by a new generation of urban terrorists. Hence, we can eliminate option (a) of the list above. Options (b) and (c) however still remain. With regard to vulnerable but not critical parts of our infrastructure, one might think of office buildings or shopping malls after having been closed for the day: as the arson attack of the first generation of the German RAF against two department stores in Frankfurt (Kaufhaus Schneider and Kaufhof) on the night of 2–3 April 1968 showed, it is fairly easy to smuggle in an incendiary device that causes great damage – but, since there was no loss of life involved, mainly for the insurance companies. According to Aust (2008, 32), it amounted to rather moderate sums, DM 282,339 (about €140,000) in the case of the Kaufhaus Schneider and DM 390,865 (approximately €195,000) for the Kaufhof. The temporary closure of these two department stores also did not affect the overall shopping behaviour of the citizens of Frankfurt who simply shopped somewhere else for the time being. An example for critical but not vulnerable parts of critical infrastructure would be the Bank of England's headquarter in Threadneedle Street, London. It is not commonly known that Mohammad Sidique Khan and Shezad Tanweer, the leading two members of the four London 7/7 bombers, were offered a list of three potential targets for their suicide attack while they were attending an Al Qaeda training camp in Afghanistan: London Underground, the Bank of England, and the G8 Summit in Gleneagles (see for example Robertson, Cruikshank and Lister 2012). Khan and Tanweer went for the London Underground straight away, reasoning that Gleneagles in Scotland was unfamiliar to them and probably also too well protected, while the headquarter of the Bank of England seemed to be impenetrable, even though the plan foresaw to drive a truck bomb into the entrance (Gardham 2012). Hence, realizing that the

Summit and the Bank of England were formidable 'hard' targets (which means, well protected ones), they went for the 'soft' target (that is, an unprotected one) in the shape of London's iconic underground mass transport system. There certainly are many more office buildings, malls and other facilities that are not critical even though they are vulnerable, and there are certainly many other office or government buildings as well as other facilities which are out of (easy) reach for terrorists. This however still leaves plenty targets in our cities which are critical, vulnerable, and under threat: airports, railway stations, busy shopping streets, shopping malls during opening times, schools, office buildings during working hours, and concert halls, museums or sports venues when they are full of people, for example. And since the attack on a Catholic church in rural Saint-Étienne-du-Rouvray on 26 July 2016 (BBC 2016), we know that even churches are no longer safe from being attacked and priests, nuns or worshippers from being murdered – but that places of worship seem to be fair game as well for terrorists, especially those affiliated to Al Qaeda or ISIS, was a lesson Shiites had to learn long before us, and Maronite and Coptic Christians in the Middle East too.

It goes without saying that it is not possible at the moment to prevent all terrorist attacks by nipping them in the bud – the incidents of Paris (13 November 2015), Brussels (22 March 2016), Nice (14 July 2016), Berlin (19 December 2016), Manchester (22 May 2017) and Barcelona (17 August 2017) are just a few of the most recent examples for the IRA's famous statement that the terrorists need to be lucky only once, while we need to be lucky all the time. Hence, the art of critical infrastructure protection is not exclusively about protection as such but also about responding to incidents, and about mitigating the consequences of an incident. This 'consequence mitigation' can be described as 'returning to normal' as quickly as possible. For example, after the devastating London 7/7 suicide bombings, most bus services were reinstated later that very day, while London Underground services resumed the next day – minus, of course, the lines where the bombings had taken place. The realization that it is well-nigh impossible to nip all terrorist threats in the bud leads to yet another issue: risk managers need to decide with asset faces the highest risk, and thus needs to be prioritized with regard to protection, response, and consequence mitigation. For example, in the context of cities, attacks on electric power substations – as the IRA had planned in the 1980s – would result in a blackout in parts of the city. Hence, protecting them is indeed important as they are vulnerable, critical, and (sometimes) under threat. However, so are obviously underground stations (London 7/7), railway stations (Madrid and Brussels), airports (Brussels, Moscow, Glasgow), hotels (Mumbai, Islamabad, Jakarta, Brighton), shopping malls (Nairobi, Stockholm), concert venues (Paris, Manchester), nightclubs (Orlando), Christmas Markets (Ludwigshafen, Berlin, Strasbourg (threat only)), pedestrian areas (Barcelona) or even beach promenades (Nice). And since, unlike substations, they are all packed with people, they need to be even higher on the priority list.

How the final list looks like very much depends on the locality in question: a small plant is easier to protect than a bigger one, and a small town easier to protect than a sprawling city. And, of course, this list would not be set in stone: any major attack will have an impact on it – the reassessment of the security of major events after the suicide attack on the Ariella Grande concert in Manchester being a case in point. One issue is clear however: there simply will not be enough security personnel and a budget high enough to protect all identified key assets with the same level of security – which implies that prioritization is quite important for allocating usually scarce resources. Furthermore, if the site to be protected is a sprawling city containing many critical assets, protecting it is an exercise that brings multiple partners to the CIP table – partners hailing from the public sector such as first responders for example, and partners hailing from the private sector such as the owners of power plants, malls or airports or hotels as well as private security companies: in Western liberal democracies in a time of privatization, most of the key assets deemed to be part of the country's or city's critical infrastructure are privately owned nowadays. Hence, on a higher than an individual facility level, a viable and successful risk management depends on a well-established public-private partnership.

This public-private partnership means no less than the need for the state and its agencies to cooperate with the private owners of those key assets that made it to the top of the priorities list – a process that is, unfortunately, less than smooth. After all, private sector ventures are basically about generating profit, mainly for the respective company's shareholders. This is not only a 'money thing' however: what also should not be forgotten in this context is the fact that risk or security managers are faced with a balancing act, even if all the money they require would be available: as Price and Forrest point out for the case of aviation security, "[security] managers must balance their efforts to protect lives and infrastructure with policies and methods that enable aviation to provide commerce with efficient and reliable service" (Price/Forrest 2016, 10). Allocating too many resources, both in terms of manpower and monies, to harden one's facility against an attack that in all probability will never happen thus does not really feature high on the to-do list. Many risk managers hence tend to adopt a rather cautious approach, on the line of 'it probably won't happen, but if it happens then not on my watch – but if it happens on my watch, it will be a one-off.' An anecdote from one of my former students illustrates this perfectly. The former student, who used to be a security officer at a major regional airport (which shall remain unnamed), reported the following conversation:

Security officer:	"I'd like to study this [university course on terrorism] – any change I could get funded?
Line manager:	"No, terrorists will never attack our airport"
* A terrorist attack occurs at the airport *	
Security officer:	"Do you fund me now?"
Line manager:	"No, they won't come back anyway…"

Unfortunately, this line manager is not the only one who prefers to shrug off the threat posed by terrorists – as already stated, many risk managers tend to think along the same lines. And this is only the tip of the iceberg. It would however be unfair just to blame the private sector for their complacency: in a time of austerity and privatization, many governments tend to either drastically reduce or to outsource crucial services without necessarily thinking through the implications for security in general and critical infrastructure protection in particular. In the third part of this book, the issue of the on-going privatization of key services of the state will be discussed in detail. Here, suffice it to say that risk management is not a luxury – the events of Paris and Brussels should have driven home that particular lesson. The USA's National Infrastructure Protection Plan of 2009 leaves no doubt of the importance of the matter:

> Risk management actions involve measures designed to: prevent, deter, and mitigate the threat; reduce vulnerability to an attack or other disaster; minimize consequences; and enable timely, efficient response and restoration in a post-event situation. (NIPP 2009, 43)

Prevention, deterrence, vulnerability reduction, consequence management and threat mitigation: these are the key parts of, and the key reasons for, a systematic approach to critical infrastructure protection in a time of complex-coordinated urban terrorism. The discussion above also implies that it is very difficult to find an answer to such complex terrorist attacks. Even in the case of rather straightforward gun assaults on just one target, like those of Lod, Vienna and Rome airports, or Chhatrapati Shivaji main station in Mumbai, police forces found it nearly impossible to adequately respond. First of all, the police officers present at the location, or reacting to an enfolding situation as first responders, are not normally trained for such situations – rather, they have probably never ever fired a shot in anger throughout their whole careers. They might also be under-equipped when compared to the arsenal the terrorists usually bring with them: in the case of the police forces at the Chhatrapati Shivaji main station, the two terrorists had AKM-type assault rifles and hand grenades, while the police officers were equipped either with simple lathis (batons) or with obsolete Lee-Enfield bolt-action rifles – a rifle type that saw first action in 1895. Hence, just two highly motivated, well-trained and well-equipped terrorists seriously outgunned them. Likewise, the police officers first on the scene at the Bataclan theatre in Paris were armed with pistols only and thus had to wait for the arrival of Police Special Forces whose members are trained and equipped to deal with such hostage crises – even though they managed to kill one of the perpetrators with a lucky shot.

But even for well-trained police special forces, responding to an on-going gun assault, usually without reliable information on the number of assailants and not necessarily with any situational awareness, neutralizing the terrorists can be quite a challenge: too many civilians might still be milling about, obscuring a clear line of fire, and also allowing terrorists to simply 'blend in' and escape after ditching their weapons – as the third Brussels Airport suicide bomber, Mohamed Abrini (also known as the 'man in the white coat' from the published CCTV stills of the attack), did after he was blown away from his own suitcase bomb due to the force of the explosion of the second suicide bomber's device (Miller 2016). Responding to a siege situation or hostage crisis like Moscow Nord-Ost or Beslan is even more daunting, all the training notwithstanding: Until the special forces required to deal with the incident arrive at the scene, terrorists have plenty time to 'dig in,' i.e. to arm doors and other obvious entry spots with explosives, and to prepare their hostages as 'human shields' or even as (involuntary) human bombs by rigging them with explosive vests or belts. This, and the fact that responding police forces are, again, usually not familiar with the location means that when the shooting starts, there is not much of a chance to save the lives of all the hostages. In both cases, gun assaults and hostage crises, first responders are on a back foot anyway insofar as they have to react to the incident. 'React' is the operative word here: even a swift reaction as was arguably the case in the Bataclan hostage crisis takes time – precious time that allows the terrorists to continue killing at will, and/or hardening the venue against a counter-attack. Not surprisingly given the manifold problem for police forces, it was only a question of time until gun assaults and hostage crises were used in combined, complex and co-ordinated attacks against multiple targets, thus overwhelming the capabilities of the police to adequately respond. After all, as already stated above, terrorist groups learn from each other.

There is a third lesson involved as well – a lesson along the lines of the famous adage 'be careful what you wish for – it might come true.' This is the realization that decentralized groups might be much tougher to fight against than centralized groups. On the one hand, systematically hunting down the known leaders, like those of Al Qaeda and ISIS, might well disrupt the chain of command, disabling them to carry out sophisticated operations in the line of 9/11. On the other hand, the multi-tiered, amorphous organization that takes the place of the centralized one is far more difficult to combat. For example, in the case of both Al Qaeda and ISIS, there is a well-known core of operators that is targeted by air raids and drone strikes – but there also is an outer ring of affiliated organisations (Boko Haram or Al Shabaab come to mind), as well as a fringe of sympathizers. Unfortunately, police and intelligence services do neither have the money nor the manpower to keep all those sympathizers under surveillance when they pop up on the radar screen – which is not always the case anyway. Hence, two of the four London 7/7 suicide bombers, Mohammed Sidique Khan and Shehzad Tanweer, could travel to Pakistan and maybe even to Afghanistan in order to train for this event without this being picked up by British intelligence services, while the perpetrators of the Madrid train bombings planned and carried out the attack without any outside assistance, also without coming to the attention of intelligence services.

Likewise, a substantial number of ISIS sympathizers are able to slip out of their home countries, take part in the civil wars in Iraq, Syria and Libya, to then return again – all of them still either so-called 'clean skins,' i.e. individuals unknown to police or intelligence services, or at least considered to be of lesser interest as compared to other, better known activists considered more dangerous on the basis of available information. Salman Abedi is a case in point: a native of Manchester, he travelled to Libya, received training in bomb-building, and then returned to Manchester where he carried out his suicide bombing attack that killed 22 people and injured several dozen others. Although he had been 'a subject of interest' by a UK intelligence service, he was not deemed to be a high priority suspect, and thus not put under close surveillance. The presence of such individuals near the very locations they intend to attack also implies that unlike the 'expeditionary terrorism' (Kilcullen 2016, 117) case of Mumbai 2008, the attacks of London, Madrid, Paris, Brussels and Manchester were partly planned and fully carried out by teams already embedded in their British, Spanish, French and Belgian societies. In contrast to 'expeditionary terrorism', Kilcullen (2016, 120) calls this 'guerrilla terrorism' – which arguably is more difficult to counter than expeditionary terrorism that involves crossing borders where, at least in theory, the perpetrators could be intercepted. Hence, Kilcullen (2016, 133), while freely admitting that he initially also was in favour of what he called 'disaggregating' Al Qaeda, points out the dangers posed by a decentralized terrorist movement:

> [A] fully disaggregated terrorist movement, with an ideology insidiously attractive to alienated and damaged people likely to act on it, combined with omnipresent social media and communications tools that hadn't even existed on 9/11, could enable a spread of terrorist violence unconstrained by time, space, money or organizational infrastructure. Add in a do-it-yourself tactical toolkit of improvised weapons and random targets, and we could be looking at an atomized, pervasive threat even harder to counter than the global insurgency it had replaced.

Coincidentally, the spokesman of ISIS, Abu Mohammed al-Adnani, backed up Kilcullen's argument on this atomized threat by encouraging so-called 'lone wolves' willing to strike on ISIS' behalf to just do so: "Do not ask for anyone's permission" (as quoted in Sengupta 2016). Kilcullen also is quite right to emphasize the role of modern technology in this regard: due to the availability of modern technology including weapons and communication on the one hand and modern means of travel on the other, today's terrorists are presented with options previously out of reach of their predecessors. Modern communication for example allows dispersed networks of terrorist groups, or individuals, to co-ordinate their attacks with a precision the anarchist terrorists of the late nineteenth century could only dream of. They can plan and to a certain extent even practise their attacks on the Internet as a virtual 'safe haven', they can keep in touch via mobile or satellite phones, and they can strike with pinpoint accuracy if need be, using global positioning systems (GPS). The same is true with regard to mobility: anarchist terrorists had to travel first by train and then by ship to move from a European destination to New York or vice versa (as in the case of anarchist terrorist Gaetano Bresci who assassinated Italian King Umberto I on 29 July 1900 in Monza) – this would have taken them days, or even weeks. Nowadays, most cities are within 24-hours reach by plane. Even more importantly, today's terrorists do not need to endanger their own lives or missions with transporting highly volatile explosives such as nitro-glycerine – and they do not need to just dream about raining death and destruction on cities from flying machines as some anarchists did: they can do so, courtesy of airliners used as flying bombs. Modern terrorists also have a wide array of weapons to choose from – quite often provided by organized crime networks (Kilcullen 2016, 206): easily concealable sub machineguns plus assault and sniper rifles instead of bolt-action ones, rocket-propelled grenades and mortars instead of crude Molotov cocktails, up-to-date Afghanistan-developed (and tested) sophisticated IED technology instead of crude roadside bombs, and in the future maybe even weapons of mass destruction instead of some dynamite. Or, if one takes a look at the still rather new 3-D printer and their future potentials, terrorists could even design and print their own weapons. Robb (2007, 8) calls this the "rise of super-empowered groups as part of a larger historical trend":

> This trend is in the process of putting ever-more-powerful technological tools and the knowledge of how to use them into an ever-increasing number of hands. Economically, that's fantastic news. This transfer of technological leverage means faster productivity growth and improvements of income. Within the context of war, however, this is dire news, because this trend dictates that technology will leverage the ability of individuals and small groups to wage war with equal alacrity.

On the lower but not necessarily less lethal end, terrorists without access to modern sophisticated weapons can easily weaponize everyday tools (for examples knifes or axes) and vehicles – such as trucks, vans, family cars or even excavators or tractors. Trucks were used in the 14 July 2016 Nice Massacre (86 fatal injuries, 434 non-fatal injuries), in the 19 December 2016 Berlin Christmas market attack (12 fatal injuries, 56 non-fatal injuries) and then again in the 7 April 2017 Stockholm attack (5 fatal injuries, 15 non-fatal ones), vans in the Nantes van attack on 22 December 2014 (one fatal injury, 10 non-fatal injuries), the London Bridge attack of 3 June 2017 (8 fatal injuries, 48 non-fatal injuries), the Finsbury Park attack on 19 June by a Far-Right sympathizer (1 fatal injury, 8 non-fatal injuries), the Barcelona attack of 17 August 2017 (14 fatal injuries, 118 non-fatal injuries) and the New York City attack on 31 October 2017 (8 fatal injuries, 11 non-fatal injuries), and family cars in the 30 April 2009 (non-Islamist) incident in Apeldoorn, Netherlands targeting the Dutch royal family on the occasion of the Queen's Day Parade (7 fatal injuries, 10 non-fatal injuries), the 20 October 2014 ramming attack in Saint Jean sur Richelieu, Canada (1 fatal injury, 1 non-fatal injury), the 21 December 2014 Dijon car attack (11 non-fatal injuries), and then again in the Westminster car attack on 22 March 2017 (5 fatal injuries, 49 non-fatal injuries). An excavator-type tractor was used in a ramming attack on 4 August 2014 in Jerusalem (1 fatal injury, 5 non-fatal injuries). In sum, what has been learned the hard way by counter-terrorism forces is that modern terrorists are far more mobile, far better informed as well as interconnected, and also far more lethal, than their predecessors ever were. Thus, at least in theory and with a lot of luck and some cooperation, modern terrorists could indeed launch complex, coordinated and very sophisticated attacks aiming at bringing our societies, and our cities, to their knees in sustained campaigns of urban guerrilla warfare. Disrupting the fabric of our societies is exactly what Osama bin Laden and Ayman al-Zawahiri hoped to achieve. Not surprisingly, the PCCIP (1997, x) reached the following conclusion: "Waiting for disaster is a dangerous strategy. Now is the time to protect our future." This is even more imperative now, two decades after this finding – also, of course, begging the question of how exactly to protect our future as well as our present.

3.2 How to Respond: The Criminal Justice Model and the War Model

At the moment, and in the foreseeable future, Al Qaeda and ISIS plus their allies pose the greatest danger to our open societies and our cities. According to Rapoport (2004), these religious terrorists belong to the 4th wave of modern terrorism – the previous three having been the anarchist terrorists (1st wave), ethno-nationalist terrorists (2nd wave), and New Left terrorists (3rd wave). Amongst terrorism scholars, this new wave sparked a debate revolving around the question whether this kind of religiously-inspired mass-casualty terrorism should be seen as an intrinsically 'new' terrorism going beyond 'old' terrorism that was usually based in one country and rather discriminate with regard to its actions. The debate has since been abandoned without reaching a consensus. But even nowadays, nearly two decades after the terror attacks of September 11, 2001, there are counter-terrorist specialists who divide the world into a pre- and post-9/11 one, arguing that ever since the world has been a different place. As Crelinsten (2009, 9) remarks on this 'new terrorism' discourse: "Before 9/11, terrorism was primarily seen as a form of crime; after 9/11 it has been transformed into a new form of warfare." If one looks at the continued global war on terrorism (GWoT), or at what Stiglitz and Bilmes call the 'Three Trillion Dollar War' (Stiglitz and Bilmes 2008), this argument is probably right. But this is not the place to reopen the debate. What is more important in our context is that the way terrorism is conceptualized also influences the way it is dealt with and reacted to: if it is regarded as a form of crime, a criminalist approach will most likely be applied – but if it is conceptualized as a kind of warfare, targeted societies are prone to retaliate in kind by deploying their armed forces, both inside and outside of one's own borders. The former approach is known as the Criminal Justice Model (CJM), the latter as the War Model (WM).

As the name implies, the CJM perceives terrorism mainly as a criminal phenomenon, basically as organized crime committed for political ends instead of for personal gains. The CJM depends on the ability of intelligence, police and the judiciary to investigate acts of terrorism, to arrest the perpetrators, and to bring them to justice. An example for the exclusive use of the CJM would be the Federal Republic of Germany (FRG) in the 1970s in its fight against the German *Red Army Faction* (RAF), also known as the *Baader-Meinhof Gang*. For constitutional reasons, turning to the Bundeswehr (German Army) for help would not have been possible then, hence the German federal and state police forces had to use their own resources, adapt their capabilities and extent their powers based on a raft of anti-terror laws swiftly approved by the Bundestag (German Parliament) in order to hunt these terrorists down. Quite importantly in this context, the German RAF conducted their terror campaigns mostly within the borders of the FRG. As such, the sole use of police forces, which only have powers of arrest within their own borders, posed no problem, while the few exceptions could be handled in cooperation with French, Dutch and Belgian police. Furthermore, the German RAF even in its heydays was a rather small organization, with only a moderate amount of firepower, and with a rather restrictive target selection that usually avoided injuring or killing innocent bystanders. Thus, drawing on the resources of the military in order to support police operations was not necessary.

Faced with the new threat of transnational terrorism of Al Qaeda and ISIS, this has changed: first, terrorists act globally, and second, they may have access to firepower that matches that of the ordinary police forces and might even overwhelm it – at least initially and at the chosen venue of attack. With regard to the latter, the commando-style attacks of Lashkar-e Tayyiba in Mumbai (November 2008) and of Daesh-affiliated cells in Paris (November 2015) and Brussels (March 2016), AK-47/AKM assault rifles, hand grenades and improvised explosive devices (IEDs) provided them with an arsenal of weapons usually associated with an infantry platoon or a special forces unit – including paramilitary armed police forces. With regard to the former, the existence of training camps on the territory of countries such as Afghanistan, Pakistan, Somalia, or Iraq creates a difficult to surmount problem for the CJM: only very few states possess paramilitary special police forces which are capable to operate outside their own country in order to hunt down terrorists – the French *Groupe d'Intervention de la Gendarmerie Nationale* (GIGN) or the German *Grenzschutzgruppe 9* (GSG9) would be an example for that – and even if they possess such a force, its deployment usually requires the permission of the government affected. Thus, temporarily operating in a (then) weak state such as Somalia might be a possibility (again, the German GSG9 would be an example), but most certainly not in the hostile current high-risk environment of Afghanistan, Iraq, Libya or Syria as this would stretch the capabilities of usually rather small police paramilitary forces a bit too far. Thus, in order to combat global terrorism on the so-called 'far shores' – or 'even at the Hindukush', as German Minister of Defence Peter Struck quipped on 4 December 2002 – as well as to prevent transnational terrorist organizations such as Al Qaeda and ISIS from sending wave after wave of well-trained terrorists to 'our shores' and our cities, deploying the military makes eminent sense: it has the superior firepower, the equipment and the logistics needed to continuously operate on foreign soil if required. Enters the War Model….

There is yet another advantage of using the military: usually, the military is vastly superior to the police, even paramilitary police forces, in terms of manpower. Bernard Loo (2005, 1) highlights this important aspect by pointing out that "Counter-terrorism is time-consuming and manpower-intensive [and there] is, hypothetically, an almost endless number of potential terrorist target sets, ranging from crowded urban centres, to high-value government installations such as power grids, symbols of national power such as government buildings and other installations." Although this remark again highlights the importance of a proper risk assessment to reduce this seemingly endless list of targets, and although it could be argued with Sun Tzu that 'he who tries to protect everything protects nothing,' it is true that in a major terrorist event, the resources even of a well-equipped police force alone might be insufficient. This is exactly the reason why after the Paris attacks, French military units supported the French police in the task of guarding critical infrastructure, and it is also the reason why well-armed Belgian soldiers were seen on patrol in the streets of Brussels. And this also explains why even in countries where the military is constitutionally banned from being used domestically, the events of 9/11 triggered debates that aimed at changing or at least amending the constitution 'just in case'. Again, Germany would be a good case in point: on 17 August 2012, the German Federal Constitutional Court ruled that in major disasters or crises, the Bundeswehr can be deployed – but minus their weapons of war, and only in assistance of the police forces. Hence, in the case of an amok run in Munich on 22 July 2016, units of the military police initially were on stand-by: in the early stages of the drawn-out event, it was deemed possible that this was an ISIS-related terrorist attack with multiple perpetrators involved. Fortunately, it quickly turned out that the Bavarian police, deploying around 2000 officers including police special forces, was capable to deal with the incident without army support.

Apart from sentry and patrol duty, there is a scenario which would involve the military as responders to domestic terrorism, and this is a terror attack using – or threatening to use – a weapon of mass destruction (WMD), be it a chemical, biological, radiological, nuclear, or high-yield conventional explosive (CBRNE) device. Although such a UHI scenario is not that likely to occur as discussed above, it is clear that if it were to happen, existing civil decontamination units, for example those of some fire-fighter departments or of other civil support institutions, would simply not be capable to deal with the aftermath of a major CBRNE attack on their own, and they might even ask military units for assistance in the case of detecting a device containing suspected chemical biological, radiological or nuclear material: not every city, county or federal state has its dedicated units specialized in dealing with hazardous material ('Hazmat'). It can thus be argued that the deployment of the military as a response to major terrorist incidents would help to shore up, or to restore, the confidence of the public in the government's ability to react to, and to contain, the crisis.

3.3 The New (Para-) Military Urbanism: Lessons from Ben Tre, Grozny and Fallujah

Making a case for the deployment of armed forces in major terrorist crises does however not necessarily mean that the military would be overly keen to get deployed to defend our cities against terrorism. This reluctance is especially pronounced if this defence would entail not only guarding critical infrastructures such as airports or railway stations as a military aid to civil powers (MACP) but also pro-actively hunting terrorist hit squads. This may sound a bit counter-intuitive at first glance – after all, pro-active combat operations are exactly what the military trains for and excels in. However, cities as well as any other heavily built-up areas constitute difficult battlespaces for modern Western armed forces more geared and trained towards manoeuvre warfare. This is even true in the case of all-out, total wars such as the 2nd World War, as the German 6th Army learned in Stalingrad between 23 August 1942 and 2 February 1943: in this pivotal battle that caused approximately two million casualties on both sides, the 6th Army as one of the Wehrmacht's most mobile and most successful combat units was annihilated. The survival of the city as such during this climactic event was irrelevant – both for the German invaders and for the Soviet defenders. Not surprisingly, most parts of the city, especially the inner city, were reduced to rubble, and about 50,000 of its inhabitants perished during the fierce house-to-house, floor-to-floor and room-to-room battle in which quarter was rarely given or expected.

Even in wars that fell short of the 'total war' assumption, cities and towns had to pay a high price for their defence or their 'liberation' from occupying forces. Too often, "it became necessary to destroy the town in order to save it," as an unnamed US Army major told CNN's Peter Arnett after the utter destruction of the South-Vietnamese provincial town Bên Tre in February 1968 (Arnett 1968). The same could have been said for the Battle of Hue at the same time and in the same conflict, or, in a different time and a different war, after the two battles for the Iraqi town of Fallujah in April 2004 (Operation Vigilant Resolve, 4 April to 1 May) and in November/December 2004 (Operation Phantom Fury, 7 November to 23 December). The latter operation was explicitly acknowledged as "some of the heaviest urban combat U.S. Marines have been involved in since the Battle of Hue City in Vietnam in 1968" (Ricks 2007, 399). It also 'became necessary to destroy the town in order to save it' in the case of the Chechen capital Grozny, which was occupied by Russian forces after a fierce battle lasting from 31 December 1994 to 8 February 1995, and then re-conquered again against Chechen insurgents in yet another fierce battle, lasting from 6 August to 20 August 1996. During the first battle, about 35,000 inhabitants were killed or injured, while in the second one, another 2000 civilians became casualties (Evangelista 2002, 44).

What renders such urban operations prohibitively costly not only in terms of infrastructure destroyed but also in terms of human lives lost is the fact that a battle in such an environment quickly dissolves into dozens, or even hundreds, of 'little battles' in the shape of merciless and ferocious engagements usually fought close-quarter-style between small-size units. Kilcullen (2016, 134) describes urban operations as follows:

> They bog armies down, breaking large units into small groups so that, instead of one big battle, there are dozens of fleeting close-range fights with a few people on either side. Most engagements last only seconds, and are fought at less than 50 yards' range. Threats can come at any moment, from any direction, including overhead or underground, and there are no safe areas. Snipers, roadside bombs, grenades, flame weapons, rockets and mortars are weapons of choice.

Hence, it comes without surprise that those civilians unfortunate enough to be trapped in their city turned battlespace frequently end up in the cross fire, and perish there. And since small combat units prefer to avoid open exposed spaces prefer to literally go through walls with the help of explosive charges (Weizman 2006)) to move through the urban space laterally and vertically from room to room, floor to floor, and house to house using a tactic called 'mouse-holing', 'Rhizome Manoeuvre' or 'infestation' (Williams and Selle 2016, 53), hiding in one's living room or even in one's cellar does not much improve the chance of survival for the beleaguered and terrified civilians either. The tactic as such is not new: Thucydides mentions that the citizenry of Plataea themselves broke down dividing walls between their dwellings to allow for a swift and unobserved movement of their own soldiers in order to isolate and defeat the 300 Theban occupation troops in their midst in the year 431 BC – the year that saw the beginning of the Peloponnesian War (Bagnall 2004, 131–132). And as mentioned above, insurgents used the tactic as well by moving "both laterally and vertically through the buildings" (Burton 1975, 3) in the few cases they were strong enough to extend mainly rural struggles into the towns and cities. In any case, the inhabitants of the Palestinian city of Gaza for example had a lot of stories to tell on this tactic after the Israeli Defence Force's (IDF) Operation *Cast Lead,* also known as the 'Gaza War' of 27 December 2008 to 18 January 2009. And still, they were probably better off than the inhabitants of Falluja or Grozny who, on top of this mouse-holing, saw whole parts of their cities utterly destroyed by a combination of air strikes and incessant artillery barrages greatly surpassing anything the IDF was throwing at Gaza.

Obviously, when it comes to defeating an enemy at the 'far shores', i.e. abroad, societies are more than willing to adopt rather drastic measures to get the mission accomplished as speedily as possible – be this mission a symmetric war against another state and its armed forces, or an asymmetric one against a guerrilla or insurgent movement. At the home front however, a similar 'no holds barred' approach leading to a 'Groznyfication' or a 'Fallujahization' of our own cities is well-nigh impossible, and it is highly unlikely that a government of a Western liberal democracy would look forward to using the 'we had to destroy the city in order to save it' defence in parliament or a media conference to justify such a desperate operation. Neither would the government of the day be keen to see main battle tanks (MBTs) rolling through the streets of their cities – even though special versions of 'urban warfare' MBTs outfitted with shorter main gun barrels, CCTVs, remotely controlled machine guns, additional protection against IEDs and bulldozer blades for the removal of barricades now are readily available: the connotations of 'losing control', 'civil war' or 'putsch' would be too strong. In any case, unleashing the proverbial 'dogs of war' in one's homeland and one's own cities would come with a hefty price tag – a price tag that would include a quick loss of sympathy within the affected population which may initially be in favour of the deployment of soldiers in the vain hope of increased security, and a swift defeat of the terrorists. Many more disadvantages connected to the use of the military in an urban setting could be discussed, for example by referring to the British armed forces' experience throughout the so-called 'Troubles' in Northern Ireland, but suffice it to say that the new 'military urbanism' is a rather reluctant one: basically all military forces of the Western liberal democracies (it may be a different situation in other political systems, but let us leave them aside in this context) firmly point in the direction of the police when it comes to combating terrorism on the home front.

Now, as already indicated above, even prior to 9/11 some Western European countries had police forces that were organized, at least in part, as paramilitary forces – paramilitary here defined as forces with a similar organizational structure as the military, but outside of the military chain of command, and usually equipped and trained like light infantry forces. Peter Kraska (2007, 6) calls them "police paramilitary units" (PPUs). The sixteen German federal states' *Bereitschaftspolizei* units for example are organized into so-called *Hundertschaften* (lit: hundredships) and, thus, very similar to light infantry companies of the army. Their arms include submachine guns, assault rifles, and machine guns, and some units also have their own armoured cars. The same can be said about the French *Gendarmerie Nationale* – except that this service is under a central, national command. Similar paramilitary forces exist in most Western European states, for example in Italy (Carabinieri) and in Spain (Guardia Civil). In all cases, these units can be deployed in support of the 'normal' police forces – not only in a terrorist incident, but whenever the situation requires more manpower than usual. Major sporting events would be one example; disasters such as floods yet another.

Other European states, for example the United Kingdom, were slower to develop similar paramilitary forces, mainly due to another tradition and understanding of policing. But since 9/11, and certainly under the impression of 7/7, even the British police forces reorganized to include armed rapid response units capable and tasked for supporting local police forces in the case of a major event. While the bulk of police officers still is unarmed, give or take the odd cans of pepper spray and Tasers, there are now specially trained firearms police officers available to back up their unarmed colleagues in a serious incident – whatever its nature. Under the impression of the Paris attacks, their numbers have been increased. The London Metropolitan Police for example states that it is now able to respond to not only one but several simultaneous Paris-style attacks, drawing on a total of 2800 firearms officers prepared to swiftly respond in a "rapid, all-out assault on attackers whose likely intent is to inflict the maximum number of casualties before sacrificing their own lives" (Milmo 2016). In the case of Scotland, about 400 firearms officers supported by armed response vehicles (ARV) are on stand-by to react to an incident (Swindon 2016). It could thus be argued that at least with regard to firepower, home-grown terrorism would also not pose a serious problem to the British police. Hence, it is fair to say that in Western Europe, the availability of paramilitary police forces heightens the threshold for the use of the military – at least somewhat, as the incidents in Paris, Brussels and Manchester indicate: in all three cases, armed soldiers were deployed to assist the police in the immediate aftermath of the events. This had nothing to do with firepower but with the fear that several terrorists would still be on the loose and ready to strike – which means that 'boots on the ground' to support the few available 'police officers on the beat' may well be the more important issue than firepower in a time of austerity. But I shall return to this issue in the third section. Be that as it may, on the other side of the Atlantic, in the USA and Canada, similar developments took place – although in the case of the USA, the National Guard was already well placed to offer support to local police.

The proliferation of paramilitary (police) forces however is only one aspect of a gradual change of modern policing: another one and quite an important one is the gradual militarization of police forces, going hand in hand with the creation of police special forces during the 1970s – in Western Europe mainly in response to the Munich Olympic Games Massacre of 1972, such as the German GSG9 and the

French *Groupe d'Intervention de la Gendarmerie Nationale* (GIGN), while the US American police special forces can be traced back to the Los Angeles Police Department's (LAPD) decision in 1974 to create a *Special Weapons and Tactics* (SWAT) team under the impression of "a series of disasters for law enforcement agencies around the country, and the general realization that traditional, person-to-person law enforcement methods just weren't capable of resolving certain issues" (Halberstadt 1994, 15). According to Kraska, militarization of the police can be defined as "the process whereby civilian police increasingly draw from, and pattern themselves around, the tenets of militarism and the military model" (Kraska 2007, 3). As Kraska further explains,

> PPUs derive their appearance, tactics, operations, weaponry, and culture to a significant extent from military special operations units [...] With BDUs, heavy weaponry, training in hostage rescue, dynamic entries into fortified buildings, and some of the latest military technology, it became clear that these squads of officers fall significantly further down the militarization continuum – culturally, organizationally, operationally, and materially – than the traditional, lone cop-on-the-beat or road-patrol officer (ibid.).

This blurring between military and police can create problems of its own – such as the (over-) use of special police forces for everyday policing or for simple riot control. In the case of the USA for example, these issues came to the fore in August 2014 in the aftermath of "the fatal police shooting [in Ferguson, Missouri] of an unarmed African-American teenager, Mike Brown, and the blatantly excessive and thuggish response to ensuing community protests from a police force that resembles an occupation army" (Greenwald 2014). Similar problems were already flagged up by Maryland's then state attorney for St. Mary's County Richard D. Fritz in 2007 when he argued on occasion of a botched raid resulting in the death of a citizen that "As State's Attorney, I can think of no greater threat to the good relationship existing in our community as it relates to police/citizen relations than to witness the unbridled use of overly aggressive tactics by a faceless and shadowy paramilitary police unit that arrives from some distant part of our state" (Fritz 2007, 15). In a sense, this is a problem of 'use it or lose it': once these special forces have been set up, initially in order to respond to the most serious manifestations of crime including terrorism, senior police leaders might become eager to 'showcase' them, even if this means deploying them in incidents that do not warrant the presence of an elite police force – or incidents where the presence of militarized police units might actually aggravate a volatile situation.

On the other hand, these 'faceless and shadowy' police special forces are indeed the units of choice when it comes to countering serious crime – especially complex-coordinate terrorist incidents as in the cases of Mumbai, Paris, and Brussels, which Kilcullen (2016, 223) calls "mobile active shooter" scenarios. In the case of the Bataclan theatre siege for example, the police special forces in the shape of a unit of the French Police Nationale's *Recherche Assistance Intervention Disuasion* (RAID, counterpart of the Gendarmerie's GIGN) and a unit of the French Ministry for Domestic Affairs' *Brigades de Recherche et d'Intervention* (BRI) reached the scene within twenty minutes, equipped not only with the expertise and the weapons required to deal with a hostage crisis but also with detailed blueprints of the building itself. This allowed for a smooth, flawless operation during which the three terrorists died, and many hostages were rescued who would otherwise have been killed by the ISIS-affiliated perpetrators. Which means, this undeniable 'militarization of police' does have its advantages in a time of urban guerrilla warfare – which by itself probably renders a further militarization of the police unavoidable. As Kilcullen (2016, 206) points out, "[thus], to critique gunned-up police forces without noticing how gunned-up the criminals are becoming is to do police officers, who operate today in a more dangerous environment […] a grave injustice." In any case, what we have here basically is a 'halfway house': it is still more about policing than about warfighting, and, hence, still more about the Criminal Justice Model (CJM) than about the War Model. In a sense, this new way of counter-terrorism policing under the threat of global, complex-coordinated terrorism could be called a hybrid CJM model.[1]

3.4 The Law: Liberal Democracies and the Limits of Response

Now that we have discussed the role of the military and the police for counter-terrorism, it is time to at least briefly acknowledge the 'J' of the CJM – justice. According to usually Roman-based Western law principles, there cannot be any crime without a law that says so (nullum crimen sine lege) nor can there be any punishment without law (nulla poena sine lege). Interestingly, prior to 9/11, many Western countries did no possess specific laws against terrorism. Germany for example brought RAF members to justice simply by focussing on the serious crimes they committed during their terrorist campaigns, such as premeditated murder, arson, aggravated assault or bank robbery. Ignoring the political background of these crimes however made the protracted court cases awkward at times: it was clear even to the most casual observer that these crimes were intrinsically different from ordinary crimes, and that the defendants were no ordinary criminals. Even more awkward was the fact that in order to combat the RAF, very wide-ranging laws were passed in a hurry that gave the police vastly extended powers. In that specific German context, this raised the question of how far we can go when it comes to eradicating terrorism. Likewise, the Italian state passed very similar laws to enable its police forces to fight against the Red Brigades as well as the mafia, resulting in the same question.

Nowadays, and as a consequence of 9/11 and the ascent of global terrorism, basically all Western liberal democracies have anti-terror laws in place, but the 'how far can we go' question still looms large. This is not just about the question of whether counter-terrorism forces are allowed to torture suspects, or whether the practice of water-boarding actually amounts to torture or not. It is, on a far more basic level,

[1] Some other terrorist scholars also are in favour of such a hybrid model, but they usually see it in terms of military aid to the civil powers (MACP), calling this an 'Enhanced CJM'.

simply about how much of our civil liberties and human rights should we sacrifice in order to be safe from terrorism: do we agree to a random 'stop-and-search' strategy adopted by the police in certain areas or not? Do we agree to the collection of our so-called 'meta data' by intelligence services or not? Do we agree to step through full-body scanners, or have our pictures taken many times by CCTV when we move through our cities or not? Or do we agree that suspects can be detained for extended periods without ever having been sentenced by a court? And, on a more general level, do we agree that special anti-terror laws are passed in order to allow the police to legally do all these things? Should we change the constitution if required to allow the use of the military for major domestic crises? And should we agree to yet another set of anti-terror laws, going beyond those passed in the immediate aftermath of 9/11? All these questions can be asked, and are actually asked. For example, in the UK, the passing of the Counter Terrorism and Security Act 2015 led to complaints of British Muslims that they were unduly targeted and discriminated against by these laws, while in the USA, the routine extension of the US Patriot Act which comes with a so-called 'sunset clause' (i.e., without another four-year extension given by the incumbent president, it will automatically expire) resulted in bitter criticism by the American Civil Liberties Union (ACLU). Whether all these vocal criticism is justified or not does not need to be discussed here. However, this debate very obviously revolves around the one crucial 'how far can we go' question: how can liberal democracies – and these are the political systems we are discussing here, not authoritarian states or outright dictatorships – defend themselves against this amorphous threat without changing into police states? Obviously, these are all important questions that need to be discussed in the broader context of our on-going war on terrorism – and, obviously, these are questions that also need to be asked again and answered in some more detail, and by way of a sober but critical debate, later on when we inspect counter-terrorism technologies as the potential silver bullets to nip the threat of terrorism in the bud.

Chapter 4
Consequences: The Urban Space as a (Limited) Battlespace

Abstract In this short concluding chapter of the first part, I bring all the argumentative strands together to then comment on the 'action-reaction' pattern that became obvious: terrorists act, counter-terrorists react – and so do politicians and academics. I argue that an array of various solutions has been offered to break through this cycle – amongst them technology as the proverbial 'silver bullet'.

Keywords Complex terrorism · Terrorism · Counter-terrorism · Counter-terrorism and technology · Academia and counter-terrorism

In the two previous chapters, we discussed the re-emergence of the urban guerrillas as predicted and theorised by Carlos Marighella, and the new military urbanism, partly in response to this development on the terrorist side. Put simply, while until rather recently, cities only ended up as a battlespace if they happened to be situated in a war zone or conflict zone, nowadays they are prone to end up as battlespace even in times of (relative) peace: they do so because terrorists go there because that is where the people are, while the military/paramilitary goes there because this is where the terrorists are. An 'action-reaction' pattern is clearly observable here: terrorists act, and counter-terrorist forces react, while interested stakeholders from politicians to private sector lobbyists to academics and journalists as the so-called 'chattering classes' comment.

4.1 Terrorist Actions

With regard to the actions of the terrorists, I demonstrated that, over the last decade starting with Mumbai 2008, the terrorist threat posed to the city evolved. First of all, terrorist groups in general are now able to launch attacks that Homer-Dixon (2009) described as "complex terrorism [that] could bring modern, high-tech societies to their knees," and that Graham (2010, xxiv) called "infrastructural terrorism." Modern technology and modern communication tools allow even small groups and 'lone wolves' to strike with a precision and a lethality rarely seen before. Indeed, as

P. Lehr, *Counter-Terrorism Technologies*, Advanced Sciences and Technologies for Security Applications, https://doi.org/10.1007/978-3-319-90924-0_4

Robb (2007, 7) warns, "technology will leverage the ability of individuals and small groups to wage war with equal alacrity." Secondly, even home-grown terrorist cells now seem to be able to strike with ever more lethality: while Mumbai 2008 – and, by extension, the Nairobi Westgate Mall siege of 21–24 September 2004 – was an act of what Kilcullen (2016) called 'expeditionary terrorism' in which a terrorist strike team crossed a national border in order to get to their chosen target, the attacks of Paris 2015 and Brussels 2016 can be seen as acts of 'guerrilla terrorism' carried out by a group living either within the city itself or somewhere nearby. Kilcullen (2016, 120–122) is quite right to argue that the latter are far less detectable for this very reason, and he is also right to point out that both Al Qaeda and ISIS gain substantial gains of a varied nature for comparatively little cost, and little risk.

On the other hand, for the counter-terrorist forces, the now rather atomized and dispersed nature of the threat renders the task of preventing an incident by nipping it in the bud more challenging than ever before: the resources of Western intelligence services simply do not allow for a constant monitoring of what are deemed to be minor figures in the Islamist scene, or for the detection of so-called clean skins before they are ready to strike. It is true that over the last decade, many sophisticated terrorist attacks were actually nipped in the bud – the Transatlantic Airliner Bombing plot of 2006 would be one example, the Ricin plot of January 2003 another. However, others such as the perpetrators of London 7/7 could successfully carry out their attack, while the London 21/7 bombers only failed due to a minor mistake in rigging their explosives, but not due to timely intelligence counter-efforts. Hence, as Paul Wilkinson once quipped, as a counter-terrorism specialist, you are similar to a famous goal keeper: people don't remember you for all your brilliant saves, but only for the one who got past you. Or as the IRA famously said, terrorists need to be lucky only once, while counter-terrorism forces need to be lucky all the time – and this is very unlikely, as we know: Paris 2015 and Brussels 2016 are but the most recent examples for the ones that got past our defences.

4.2 Counter-Terrorist Reactions

If a terrorist cell remains undetected long enough to carry out their planned attack, all we can do is to react to it. With regard to Mumbai 2008, the police countermeasures were hampered by unclear chains of command, obsolete equipment in the hands of police forces already at the scene, a lack of knowledge of the venues that were under attack, and a misguided police special forces philosophy that focused on equipment to the detriment of actual training. An aggravating factor most certainly was the death of the commander of the Mumbai anti-terrorist squad (ATS), Hemant Karkare, killed early during this incident in a firefight with the two cell members who had just attacked the Chhatrapati Shivaji train station and were at that time trying to get into the Cama hospital, Ismail Khan and Ajmal Kasab. Compared to Mumbai, the response of the special police forces in Paris was very rapid indeed – and their counter-attack at the Bataclan theatre was greatly helped by the fact that they had the blueprints of basically all major venues in Paris at their disposal. Nevertheless, the fact remains that if an attack is carried out, the response of police forces is a reactive one – which allows the terrorists to keep killing for a substantial amount of time.

'Reactive' also is the operative word when it comes to legal counter-measures. After each major terrorist incident, a number of legal initiatives are usually passed, most of them either granting more powers to intelligence services and/or the police, or plugging some newly identified chinks in our defences (Kilcullen 2016, 119). One of the main drivers in this regard was, hardly surprisingly, the events of 9/11, as the 9/11 Commission Report made clear. After having pointed out that border security "encompassing travel, entry and immigration" had not been taken seriously or deemed to be a security matter prior to 9/11 despite the fact that "more than 500 million people annually cross U.S. borders at legal entry points [while another] 500,000 or more enter illegally without inspection across America's thousands of miles of land borders", the authors of the report recommended that from now on, the "United States should combine terrorist travel intelligence, operations, and law enforcement in a strategy to intercept terrorists, find terrorist travel facilitators, and constrain terrorist mobility" (9/11 Commission Report 2004, 383–384). When the final report was published in 2004, these recommendations had already been acted upon, for example via the Aviation and Transportation Security Act of 19 November 2001 that introduced metal detectors, 100 percent baggage screening, reinforced cockpit doors for airliners and an expansion of the Federal Air Marshal Service. But that was only the start of a raft of new security measures, most of them revolving around new security technologies (see for example Peterson 2016). Frequent air travellers know this all too well: the shoe bombing plot of December 2001 for example resulted in the requirement to put one's shoes in a tray as well on request, while the Transatlantic Airliner Bombing plot of 2006 led to a restriction with regard to the amount of liquids that can be taken on board of a plane – liquids that have to be put in a transparent bag and placed in a tray on top of that. Experience also shows that after a certain amount of time has passed, security measures tend to get less strict again (to the great relieve of frequent flyers in the case of aviation security). Anderson (2008, 228) calls this the 'politics of the latest outrage:' politics in reaction to overwhelming pressure from the public and the media, clamouring that 'something needs to be done' (on aviation security in this regard, see Argomaniz and Lehr 2016).

4.3 Academic Reactions

The 'politics of the latest outrage' more often than not also result in a flurry of research activities that aim at a better understanding of the underlying causes of political extremism and terrorism. Again, this is nothing new – for example, attempts to discover a 'terrorist mind' in the shape of psycho-pathological (and measurable) personal traits can be traced back to the second half of the nineteenth century and the birth of modern terrorism. As late as the 1980s, the Ministry for Domestic Affairs of the Federal Republic of Germany funded an ambitious programme that also looked into the personal histories of the then-known Left-Wing terrorists in the hope to discover common traits, and to better understand group processes. With regard to the common traits, these hopes have been dashed time and again: there are simply no such commonalities. Today, psycho-pathological studies have been sidelined in favour of the study of radicalization and deradicalization processes. The big questions asked here are why and how do individuals become involved in political extremism and terrorism, and how do they extricate themselves (if they ever do)? The hope is to discover common and observable 'milestones' that individuals pass on their journey into terrorism – milestones that could be used for a timely intervention to prevent certain individual's, or whole group's radicalisation.

From a purely scientific and academic perspective, studying the root causes of terrorism as well as the processes that lead individuals as well as whole groups into terrorism are quite illuminating: our understanding of political extremism and terrorism certainly has improved in the last decades. Whether this improved understanding also leads to tangible results for our current counter-terrorism efforts is however questionable. For example, when it comes to the 'milestones' I just mentioned, basically all of them could be detected only retrospectively when scrutinizing the careers of terrorists in investigations started after a successful attack – for example the one reconstructing the lives of the London 7/7 bombers. The vaunted 'root causes' of terrorism (such as relative deprivation, inequality, injustice, colonial history etc.) on the other hand may well be detected but usually are woven into the fabric of our societies and their histories – which arguably makes it rather challenging to address them in the here and now, to say the least (see for example Kaarthikeyan 2005, 131; Laqueur 1987, 6–7). Not surprisingly, even some eminent terrorism scholars are of the opinion that studies aiming at a better understanding of terrorism are at best a waste of time, or an exercise in relativizing and apologetics at worst.

Personally, I do not share the often-expressed view that all we need to do is either to kill the terrorists or to lock them away. As long as we do not understand 'what makes them tick,' our counter-terrorism efforts are akin to medics addressing symptoms without curing the disease: since the cause still is there, a new generation of terrorists will carry on fighting when the original generation is either dead or in prison. European examples for this would be the Provisional IRA in Northern Ireland, the three generations of the RAF terrorists in Germany, or the Basque terrorist organization Euskadi ta Askatasuna (ETA). And on the Salafist-Jihadist side, al Qaeda undeniably went through several generations of Jihadists since the 1990s with no end in sight, while ISIS's success in recruiting disenfranchised youths usually in their late teens or early twenties also makes it very likely that we are faced by more than one generation of their fighters as well.

4.4 Looking for a Silver Bullet

I readily acknowledge that while we conduct our painstaking researches for the root causes of terrorism (as we should), people keep getting victimised by terrorists – not only on the 'far shores' such as Afghanistan, Pakistan or Iraq any longer, but also in seemingly increasing frequency in our own homelands, and our own cities. Hence, it is only logical that we are looking for silver bullets that help us to escape the aforementioned 'action-reaction' cycle – silver bullets that help us to prevent terrorist incidents by nipping them in the bud without requiring us to transform our societies into police states in which one half of the population spies on the other half. After all, in our inter-connected, globalized world, it is well-nigh impossible to "pull up the drawbridge, disengage from the world, and somehow avoid the fight" (Kilcullen 2016, 201). There is one very telling remark from Kilcullen on that our current war on Daesh and Al Qaeda's terror:

> [This] war truly is, as many have argued, a Long War. There is no magic bullet, no instant solution, let alone some carefully calibrated combination of firepower, diplomacy and technology that can quickly put the genie back in the bottle. [...] Even as of early 2016, there are still more than 35,000 fighters in ISIS, and roughly as many in the Taliban and other extremist movements. The rise of ISIS, the stimulating effect of its rivalry with al-Qaeda, the Taliban resurgence and, above all, Baghdadi's declaration of the caliphate, is breathing new life into a global movement that seemed to be fading – proving that the ideology, like the movement defined by it, is tough and resilient. This conflict will not be going away any time soon, and it certainly won't end quickly or cleanly. (Kilcullen 2016, 200)

For the quagmires of the Near and Middle East and West Asia, that may well be the case. Here, diplomacy plays a major role to bring a huge variety of allies and potential allies as well as current enemies who might end up as future friends to the negotiation table. At the home front, a plethora of more often than not ill-defined and ill-thought through preventative measures such as deradicalization programmes and grassroots community policing initiatives take the place of diplomacy – with the roles of firepower and technology remaining the same. In any case, since the circumstances of disenfranchised communities in our midst arguably vary from city to city and country to country,[1] bespoke approaches and initiatives aimed at preventing radicalization have to be found – very obviously here, a 'one size fits all' approach cannot possibly be used. The role of firepower was discussed already, and I argued that there is no lack of it. The problem rather is the largely reactive nature of our responses in the face of terrorists being the 'fast moving target' as mentioned above. Hence, Robb (2007, 153) is quite right when he argues:

> To build a solution, we need to start with the assumption that we don't know what the next threat will be. It's clear that new threats will be both amorphous and extremely difficult (nay, impossible) to fully anticipate. Our vulnerabilities are too vast and the sources of potential threat too wide to plan specific defences against.

[1] Kilcullen (2016:204) makes this argument specifically for Muslim communities.

Just how amorphous and extremely difficult to anticipate the threats posed by terrorists are became evident for example in the case of the already mentioned 14 July 2016 Nice Massacre, in which the ‘weapon’ used was a 20-tonne refrigerated truck. And a couple of days later, on 18 July 2016, a self-radicalized attacker used an axe and some knifes to seriously injure 15 people on a regional train near Würzburg, Germany. Whether such attacks actually qualify as terrorism or whether they should rather be categorized as amok runs carried out by unhinged individuals who try to ‘legitimize’ and justify their activities by superficially subscribing to an extremist ideology that basically comes in as an afterthought is a moot point in our context. More to the point, in the wake of the train attack, the Bavarian Minister for Domestic Affairs, Joachim Herrmann, correctly pointed out that even the most extensive and wide-reaching concept for security would never be able to completely prevent incidents such as in Nice and Würzburg: after all, the means of attack were everyday tools readily accessible to all citizens – tools that, unlike firearms, cannot possibly be eliminated from modern society (Munzinger 2016). Herrmann’s argumentation was mirrored by that of former extremist Adam Deen, who commented on the Nice attack as follows:

> As President Hollande termed it, the “terrorism of opportunity” is on the rise and we are likely to see more attacks of this kind. The shift away from centralised planning of attacks towards operational independence demands a nuanced counter-strategy beyond extreme state security measures. French security forces are already stretched to the limit and have been operating around the clock across the entire country. But focusing on intelligence and sharp-end security will not do the job. It is neither feasible nor desirable to monitor all individuals at risk of radicalisation or to ban individuals from renting a lorry. (Deen 2016)

In order to counter this threat, Deen recommends a new interpretation of Islam filled with enlightenment values such as liberty, equality and fraternity (Deen 2016) – an interpretation known as European Islam or *Euro Islam* (a term coined by Tibi 1995). As commendable as this is, such a new interpretation will take time to be developed and spread – and it will most certainly face an uphill battle against the already existing extremist interpretations of Islam. The same can be said for any other deradicalization programme that aims at countering and defeating extremist messages, whatever their flavour. In the meantime, and working under the assumption that such programmes might, eventually, be successful, some other approach is required that could help preventing acts of terrorism in the here and now. As the German Ministry of Domestic Affairs (Bundesministerium des Innern, BMI) stated under the impression of the terrorist attacks of Würzburg and the suicide bombing in Ansbach on 24 July 2016, we must make it as difficult as possible for the terrorists to find any targets for their attacks and to translate their plans into reality – which means we must reduce our vulnerabilities and protect our critical infrastructures (BMI 2016). This is also the position of the European Union, which, under the impression of the attacks on various European cities promised to “provide more than €100 million ($118 million) by mid-2018 to support cross-border projects and protection of the so-called soft targets in cities: public spaces with a high concentration of people” (Eubulletin 2017). Seen from a British perspective, this is not necessarily a new revelation: in 2007, and still reeling under the impact of the

7/7 bombings, HM Government inaugurated a new strategy aiming at exactly that – at protecting our infrastructures by harnessing, amongst others, the power of modern technology. The strategy in question is known as *Security and Counter-Terrorism Science and Innovation Strategy* of 2007 – abbreviated as INSTINCT. It covers both readily available commercial off-the-shelf (COTS) products as well as tailor-made solutions developed in think tanks of universities or private sector providers.

And this is precisely the reason why we now turn to these modern, sophisticated technologies either already available off the shelf, or currently being developed to counter terrorism – technologies such as biometrics, smart CCTVs, or devices that can detect explosives in particular or any odd behaviour in general from afar, to name but a few. After all, if terrorists can benefit from available technology as discussed, so can we – and if there is any magical or silver bullet in sight, it probably is modern technology that could allow us to set up an early warning and rapid response system. For the time being, and until peace is re-established in the regions affected by the 'long war', this seems to be our only chance to protect our open societies in general, and our cities in particular, from waves of infrastructural, complex-co-ordinated terrorism…

Part II
Our Technology Will Win – The Role of Technology in Urban Counter-Terrorism

After having discussed the action-reaction dynamics between terrorists and counter-terrorists in the first part, I will now turn attention to technology as the hoped-for 'silver bullet' in our fight against terrorism – not necessarily in the sense of letting us win this war (if this is at all possible), but in the sense of adding a last ditch layered defensive line under the assumption that however good our efforts are, there always will be the one that slips through the net. Although we cannot possibly stop and search everybody walking through our streets, we still need to prevent terrorists from launching an attack, or at least to minimize its consequences. Firstly, we need to make sure that a potential terrorist does not infiltrate a key asset by establishing whether he or she has the right credentials and the right intentions. Secondly, we need to detect any dangerous substances carried by an individual before he or she reaches the planned location of an attack. And thirdly, we need to have measures in place that mitigate the consequences of an attack if everything else fails. Technology can greatly help us to do all that – this at least is the opinion of the more optimistically-inclined of those working on counter-terrorism. The implications of technology are nicely illustrated in Mills and Huber's thought-provoking article, published a year after 9/11:

> "In the post-September 11 world, we know we have to see the plastic explosives in the truck before they detonate, the anthrax before it's dispersed, the sarin nerve gas before it gets into the air-conditioning duct – and not just see but recognize it. Our imaging systems will have to distinguish between the scaffolding on a Wonder Bra and the wiring on a bomb. We have, of course, slow, conventional ways of doing all this in a forensic laboratory. But the challenge now is to do it in bulk and in real time. How do you even begin to do this?" (Mills and Huber 2002, 1-2)

Acknowledging the magnitude of the challenge, they nevertheless were rather optimistic that advanced and sophisticated technology would eventually be able to defeat terrorism, even concluding their article with the claim that "[our] silicon will win" (Mills and Huber 2002, 13). In a nutshell, both authors see high-tech harnessed for the purpose of counter-terrorism as the proverbial 'silver bullet' with which not necessarily to defeat terrorism as such, but rather terrorists attempting to attack us at the home front. In a sense, and even though they also discuss an 'offensive' use of technology, their article mainly is about a last line of defence if and when terrorists managed to slip through our first (Prevent) and second (Pursue) lines of defence, and should thus be taken seriously. This requires a close look at such technologies, both as they exist as well as they currently are under development, and to critically inspect them – after all, not everybody is convinced that 'our silicon will win.' David Lyon for example is certain that when it comes to the 'technology versus terrorism' equation, "technology won't save us" (Lyon 2004, 311). It should be added that in their article, Mills and Huber solely focus on silicon-based 'digital wizardry' when discussing how counter-terrorism technology could render our open societies and our cities safer. In the following however, I will use a broader understanding of 'technology', following the Oxford Online Dictionary that defines it as the "application of scientific knowledge for practical purposes." The term itself comes from Greek *tekhne* meaning 'art' or 'craft', while the composite of *tekhne* and *logia,* i.e. *tekhnologia,* simply means 'systematic treatment.' Understanding counter-terrorism technology in the broader sense of a 'systematic treatment' enables us to move beyond the usual 'smart' array of observation cameras, biometric chips, or x-ray scanners and the like to also include passive defences in the shape of barricades (i.e. counter-intrusion devices) and citadels (i.e. facilities hardened against terrorist or criminal attacks). But let us start with what Mills and Huber (2002, 1) call the "digital wizardry."

Chapter 5
Identification: Biometrics, or a Real-Time 'Who Is Who'

Abstract In this first chapter of the book's second part, I argue that the best way to deal with terrorism is to nip a planned terrorist attack in the bud. One way to do so is by discerning innocent citizens who have a right to be at certain locations from those who have not, and whose intentions are dubious, via a swift and reliable but non-intrusive identification and verification process. I introduce the main biometrics that are used in this regard, first discussing physical biometrics such as facial recognition, iris scans, voice scans or fingerprints, to then move on to behavioural ones such as one's gait. Since most of them are not (yet) fool proof, especially not as stand-alone technologies, I make the case for multimodal-based biometric systems which currently are under development.

Keywords Biometrics · Biometric passports · Facialr recognition · Iris scans · Fingerprints · Voice recognition · Hand geometry · Palm prints · Gait recognition · DNA recognition · Multimodal systems · False positives

As already mentioned, the best way to deal with terrorism is to nip a planned terrorist attack in the bud before it can be carried out. One way to do so is by correctly distinguishing innocent citizens who have a right to be at certain locations from those who have not, and whose intentions are dubious, via a swift and reliable identification and verification process – preferably one which is not too intrusive for the general public, or one that could even be beneficial for them, for instance by doing away with the tedious duty to remember ever more complicated passwords for all those gadgets and applications we use.

The need, or rather requirement, to identify oneself seems to be one of the hallmarks of modern life: if we want to get money out of an ATM, we are required to enter a four-digit personal identification number, or PIN for short. That's perfectly fine if we have just one or two debit or credit cards that we frequently use, especially so since most banks nowadays offer the choice to change their computer-generated PIN to a number more to our liking – as long as it is not 1234 or 0000 or something similarly easy to predict. But what if we have a couple of credit cards that we use only rarely, for example to check into hotels during a business trip or while on vacations, or to hire a car? The same can be said for computers, tablets and smart phones,

P. Lehr, *Counter-Terrorism Technologies*, Advanced Sciences and Technologies for Security Applications, https://doi.org/10.1007/978-3-319-90924-0_5

while many websites and apps demand a log-in as well, which again depends on the correct entry of a password, often followed by the additional step of sending a four-digit security code send to one of our trusted devices to render computer fraud in the shape of hacking and phishing more difficult. Sure, there are password managers that we can install on all our devices, but at least for me as a frequent traveller, there always seems to be the moment when I sit in my hotel room or an airport lounge trying to log in to one of my various accounts – just to realize that this particular password isn't in the password manager, and that the envelope I scribbled it on is in the upper-most drawer of the cupboard next to my desk in my now far-away office. Wouldn't it be just great to have an alternative option when it comes to securing access to your accounts, apps, credit cards, and bank accounts? Fortunately, these alternatives already exist in the shape of biometrics.

Biometrics – a term coined from the Greek words 'bios' (life) and 'metron' (measurement) and thus means 'measurement of life' – can be defined as 'the statistical use and analysis of (unalterable) biological data' or "the use of data extracted from the body" (Lyon 2004, 299) under the assumption that everybody possesses certain unalterable features that are unique to him or her, depending on the biometrical modality in question (see for example Al-Abed, Charrier and Rosenberger 2012). The best-known ones, and the ones still most widely used, are faces and fingerprints. Everybody has their (fairly) unique face as well as a unique fingerprint pattern (even those of identical twins differ slightly), as the police well know – which is why they draw on both to identify criminals since the late 19th century. The wave of anarchist terrorism that commenced in the late 19th century as well and cost many eminent politicians their lives, amongst them Tsar Alexander II of Russia, King Umberto I of Italy and US President McKinley, accelerated the use of these identification techniques. As a result, and long before the time of computerized data bases, a first crude attempt of facial and bodily recognition was made to identify suspected terrorists across borders during the war on anarchist terrorism as the 'real' first war on terrorism. This was the so-called *portrait parlé* – according to Deflem "a method of identification that classified criminal suspects on the basis of measurements of parts of their head and body" (Deflem 2005, 278), adopted at the end of the Conference of Rome for the Social Defence Against Anarchism (24 November to 21 December 1989, attended by 54 delegates from 21 European countries).

In our time of international terrorism on the one hand and large-scale migration on the other, the usefulness of biometrics for identifying individuals did not escape various immigration authorities and counter-terrorism specialists. For example, individuals who travel to the USA as non-US citizens now are fingerprinted on arrival. This is also the reason why current versions of passports, now known as 'biometric passports', not only contain their holders' photo but also their unique pattern of fingerprints, digitally taken when an application for a passport is made, or an existing one renewed. This practice was introduced in the aftermath of 9/11 to assist immigration authorities as well as police forces in keeping track of individual travellers as well as to render the forging of travel documents more difficult. In a nutshell, as Lyon explains, these biometric passports "have been consistently touted as a key means of enhancing security – a way of being sure that people are who they say they are and that they have a right or a reason to be where they are" (Lyon 2004, 301). Indeed, when it comes to enhanced security in general and the prevention of terrorism in particular, the introduction of biometrics in travel documents can be seen as a "crucial milestone" (Malčik and Drahansky 2012, 1).

5.1 Facial Recognition and Iris-Scans: The Proverbial 'Eyes of the Beholder'

Let us start by looking at biometric passports and biometric ID cards first. These documents feature a laminated data page that contains the document holder's photo taken as a greyscale image, and a machine-readable zone (MRZ) on the bottom with basic information such as the holder's name, date of birth, place of birth, sex, passport number and date of expiry. It also includes a so-called 'passive RFID tag', that is a small ROM microchip based on radio-frequency identification (that's what RFID stands for) readable by a short-range (maximal 15 cm to prevent data intercept) RFID reader via an antenna which is also integrated in the data page. The read-only memory (ROM) contains several data groups from DG1 to DG16, but for identification purposes, only DG1, DG2, DG3 and DG5 are relevant as this is where data relevant for identification is stored: DG1 contains the same data as that of the MRZ, DG2 contains the data pertaining to the encoded face, DG3 contains the data of the encoded fingers, and DG5 contains the photo of the holder that is also on the data page. DG4 is reserved for data of encoded eye patterns, while DG6 is reserved for data of other biometrics that may be added in the future (all information from Malčik and Drahansky 2012, 2–3).

Regarding the document holder's photo, the first biometric ever used for identification was facial recognition – in the early days simply in the shape of a purely 'visual' recognition by the immigration or police officer inspecting the old paper passport or paper ID at the immigration desk, comparing the picture on the document with the person holding it. Computer-based facial recognition emerged during the 1980s, promising many improvements for security and other purposes that sounded quite appealing. As Kyle Chayka enthuses in his otherwise rather critical article on face recognition:

> Walk up to the international check-in at a German airport, gaze up at a camera and walk into the country without ever needing to pull out a passport – your image is on file, the camera knows who you are. Wander into a retail store and be greeted with personalised product suggestions – the store's network has a record of what you bought last time. Facebook already uses face recognition to recommend which friends to tag in your photos. (Chakya 2014)

Indeed, 'face prints' can be quite convenient – I shall come to the 'dark side' flagged up by Chakya as well later on. To dwell on the positive aspects first, being able to amble across an international border or into a store and immediately being recognised in a 'Minority Report'-style identification process without having to whip out a passport or an ID, or having to talk to someone, certainly has some appeal. As Malčik and Drahansky explain, this is made possible by the fact that "the face contains information that is invariant in time and can be measured [...], for example, the distance between eyes, position of chin, position of nose, and so forth" (Malčik and Drahansky 2012, 3). These features allow a computer to generate a statistical analysis of one's face based on them, and to compute an algorithm ('set of instructions'), 'template' or 'mathematical map' unique to the passport or ID-card holder. The picture data of about 15 kB size then is stored twice (encoded and as greyscale) on the biometric passports' microchip as DG2 (Malčik and Drahansky 2012, 3–4). Compared to fingerprints however, facial features are somewhat less 'unique' as well as less 'unalterable': for example, a hood pulled deep down on the face, a pair of polarized sunglasses or a beard or moustache (or shaving off an existing one) could probably mislead a police officer, but it can even more easily fool older computerized identification systems such as security cameras and CCTV. Hence, even today, individuals are not allowed to wear sunglasses when their passport or ID card picture is taken, or to smile: a huge grin or a guffaw would be enough to alter this algorithm, and thus defeat the purpose of this exercise. Those who travel to some of the world's major airports know that so-called *electronic gates* or *e-gates* for short make use of optical data recognition (OCR) software in combination with a passport scanner reading the data page and the passport's electronic chip to let travellers through the border – or not: since the system is not failsafe, and since people tend to forget to remove sunglasses or not to smile, the queues on e-gates can actually be longer as the ones at the 'normal' human-operated immigration desks.

Furthermore, as Introna and Woods explain, the way these 'templates' are created also leads to issues such as false positives. Basically, there are two broad approaches when it comes to digitising face images for identification purposes: image-based algorithms, which "use a template-based method to calculate the correlation between a face and one or more templates to estimate the face identity"; and geometry feature-based algorithms, which "capture the local facial features and their geometric relationship" (Introna and Wood 2004, 185). Both approaches condense the data thus derived into an algorithm of just a couple of kilobytes – after all, the microchips of biometric passports are very small when it comes to memory space. This data reduction however negatively affects the performance of the algorithms, as Introna and Wood argue. With regard to template-based algorithms, they point out that biases invariably tend to be built into the template: "It obviously depends on the gallery used to create the standard template as well as the range of potential variations within a population. For example, because minorities tend to deviate the most from the standard template they might become easier to recognise" (Introna and Woods 2004, 186). With regard to feature-based algorithms, they explain that "because of the reduction the 'face prints' generated are in close proximity to each other [...] the discrimination required for the recognition task [is made] more difficult" (ibid.).

The foreseeable consequence of this could be a flurry of 'false positives' triggered by the facial recognition systems – and, as Introna and Wood's research indicate, biased ones at that: "with an increased threshold small differences in identifiability [...] will mean that those that are easier to identify by algorithms (African-Americans, Asians, dark skinned persons and older people) will have a greater probability of triggering the alarm [which] also means that these groups would be subjected to a higher probability of scrutiny as false positives, i.e. mistaken identity" (Introna and Wood 2004, 192). From a security perspective, this is less problematic than from a civil liberties' one (I shall deal with this issue later on): after all, a 'false positive' (also known as 'false rejection' or 'false non-match') can be resolved by an immigration or police officer via questioning, while a 'false negative' (also known as 'false acceptance' or 'false match') would mean that the person mistakenly accepted by the system would be able to enter the country, or a facility, without being challenged – with the result that a chance to nip an impeding attack in the bud would be wasted. In any case, it can be said that at the moment, both kinds of algorithms obviously tend to work best "with a small database and good quality face capture" (ibid.). Although the latter can be ascertained to a certain degree if and when human operators assist in taking face captures (or at least reject suboptimal images in case of government officials responsible for passport or ID card applications), the requirement of keeping the database limited cannot be met when it comes to biometric passports or IDs. This notwithstanding, facial recognition techniques are already in use with immigration authorities of a number of countries: when entering the country, a picture is taken at the immigration desk and saved together with the details the passenger provided on the immigration form or landing card. When leaving the country again, this picture is compared to the one taken at passport control to ensure that the person leaving the country is identical to the one entering it. To sum it up, when it comes to crossing e-borders, or to gain access to certain types of critical facilities, facial recognition by and large works reliably enough, occasional mismatches notwithstanding.

Regarding non-official identification and verification procedures, the practice of using facial identification for the purpose of identifying the rightful owner has just been rolled out in the meantime on that front as well and also works quite reliably. Current examples are Apple's iPhone X and Samsung's Galaxy S9. As concerns smart phones in general, the penchant of mainly younger users of smart phones to take 'selfies' has also been turned into a way of identifying oneself, for example when it comes to paying small bills. This so-called 'Selfie-Pay' technology requires users to hold the camera of a smart phone equipped with this application ('app' for short) in front of their face. The app will then use its in-build facial recognition software to compare the facial features detected by the camera with the saved algorithm. To prevent fraud, for example using a photograph, users are also required to blink during the process to convince the advanced software that they are a 'real' person. Furthermore, facial recognition is about to be used by private companies as well, for example by some airlines such as JetBlue and Delta (both US companies). In these cases, the passenger's face will be scanned, and its algorithm compared with the one of the picture in the ID or passport. It is hoped that this could render boarding passes superfluous (JetBlue) or speed up baggage drop (Delta). Arguably, this has less to do with security than with convenience, and how fail-safe it will be remains to be seen. The same can be said for the rolling-out of facial recognition technology for Thailand's 11,000 7-Eleven convenience stores. Here, the intention is already very near to Hollywood blockbuster *Minority Report:* apart from monitoring stock levels and the behaviour of staff, loyal customers are meant to be identified, overall shopping patterns analysed, and suggestions made to loyal customers on the basis of this information (ThaiTech 2018).

There is however an even better (and also harder to fake) way to identify oneself while using a particular part of one's facial feature, and that is the iris, which is as unique as one's finger prints. Although some computer software such as Windows 10 or smart phones such as the Samsung Galaxy S9 already include the possibility of using an iris scan for self-identification, this method is currently mostly used in advanced security systems that use mathematical pattern recognition software similar to those used for finger print scanners. The system usually is faster than less sophisticated first generation facial recognition systems, and also less prone to misidentification. It is however not entirely secure as a stand-alone solution. For instance, high-quality photographs of irises allegedly could fool some of the earlier iris scanners, while other scanners turned out to be difficult to adjust from user to user – again leading to delays at the e-gates equipped with them. The fraud potential could be minimized by using iris scanners, at least for official purposes such as e-gates, in conjunction with other biometrics such as fingerprints, as is currently the case for systems using facial recognition. Problems posed by difficulties to adjust the scanners in a way that they are accessible for tall and small users alike are mostly solved by now due to technical improvements. The reserved data group DG4 for encoded eyes on the microchip hints at a possible introduction of iris recognition in the future, and so does DG6 which currently is 'reserved for future use', also leaving the door open to include other biometrics which, at the moment, are not yet ripe for being used in biometric passports.

5.2 Fingerprints: Tented Arches, Plain Arches and the Central Pocket Loop

Due to the limitations of facial recognition systems, modern biometric passports also include a second biometric feature as a fall-back option, and this is a digitised version of one's fingerprints. Fingerprints (or rather, an algorithm based on the individual pattern of one's fingerprint) were the second biometrical feature added to modern passports (DG3 on the chip): due to the unique pattern of one's fingerprint, especially the 'tented arch', the 'plain arch' and the 'central pocket loop', fingerprints have been used for forensic purposes since the late nineteenth century. For those of us belonging to the older generation, fingerprints probably are a somewhat odious way of identifying oneself as the practice of taking fingerprints is still too much associated with police work: previously, fingerprints were mainly taken (manually, not digitally) to identify criminals, or those suspected of having committed a crime. Not surprisingly, the decision to include fingerprint patterns in the current biometric passports initially met with some resistance from civil liberty groups: in March 2008 for example, the British Airports Authority (BAA) decided to remove an already installed fingerprint reading system from its new Terminal 5 at London Heathrow International Airport the day before it opened due to protests (Boussadia 2009, 7). But from the perspective of immigration and police authorities, the combination of face recognition with fingerprints simply is too valuable to not to use – especially so since fingerprints are harder to counterfeit than faces. When entering the USA for example, submitting digital fingerprints is required nowadays for this very reason.

Although faking fingerprints is by no means impossible (there are even 'how to do' instructions on the Internet), this is quite complicated at least with regard to official use: as Malčik and Drahansky (2012, 4) point out, "the fraudsters have to face the problems with tricking the fingerprint scanners, because the scanners are being more often equipped with sophisticated liveness detection – especially when a security risk is expected." This 'liveness detection' can be accomplished using a variety of methods, for example the detection of perspiration, temperature, pressure stimulus and others (ibid). Since fingerprint pattern detection as an additional identification method arguably enhances security, citizens of most Western liberal democracies now have to leave fingerprints as well when applying for a new passport or extending an old one. A fingerprint sensor digitizes the prints according to various standards and turns them into compressed greyscale images of about 15 kB size, which are then stored on the new passport's microchip (Malčik and Drahansky 2012, 5). So far, the combination of fingerprints and facial recognition in biometric passports is the state of the art – but future passports and IDs are likely to contain iris recognition features as well.

When it comes to official identification purposes, mobile fingerprint scanners already are available for about a decade now. One could even quip that 'there's an app for that' since the earliest of such system was the rather pricey iPhone app (US$3000) known as *Mobile Offender Recognition and Identification System* or MORIS for short. It was (and still is) only available for US police officers. According to the developer of the software, BI2 Technologies, MORIS "gives you [the police officer] the wireless ability to Positively Identify offenders enrolled in your local or BI2's national database in seconds – from any location where there is cell service supporting Internet connectivity" (BI2 Technologies 2009). MORIS actually does so via facial recognition, iris recognition, and fingerprint recognition, thus combining the three distinctive biometric features already discussed to ensure a correct identification. It also comes with access to SORIS and I.R.I.S., the former being the "Sex Offender Registry and Information System", the latter the "Inmate Recognition and Identification System" (ibid.). Hence, the BI2 Technologies brochure enthuses, MORIS "is ideal for identifying previously enrolled […] Sex Offenders, Illegal Immigrants, Gang Members, Individuals with Outstanding Warrants [and] Parolees and Probationers" (ibid.). In order to function as advertised, a connection to the Internet is required to access databases containing the relevant data – which already highlights the importance of the less visible (in comparison to police patrols) 'back offices' where the servers are kept. It took other police forces a while to also 'discover' the possibilities offered by modern technology, but currently, software solutions similar to MORIS are gradually being rolled out outside of the USA. In February 2018 for example, West Yorkshire Police began to test a mobile fingerprint scanner that, like MORIS, depends on being connected to a smartphone with an Internet connection to access a newly set-up database named *Biometric Services Gateway* (Higgens 2018). As regards private identification and verification procedures, (former) high-end smart phones such as Apple's iPhone 5 also took to fingerprint scanners in order to simplify the process of unlocking one's phone while also rendering it safer than the 'pattern drawing' procedure or even the four-digits pin code. But, as already mentioned above, current high-end smart phones already include facial recognition or even iris scans as even safer methods of unlocking one's phone. On that front at least, fingerprints are already 'old' biometrics, and probably retained mostly as fall-back options.

5.3 From Voice to Gait: Other Forms of Biometrics

So far, one's facial features and fingerprints are already routinely used for identification purposes, and iris recognition is not far behind. Another obvious feature that could be used for biometrical purposes is one's voice, which also is a unique personal trait since it is very difficult to fake – which is why voice recognition (or speaker recognition) is frequently used for forensic purposes such as identifying individuals in data derived from wiretapping during criminal investigations and/or in intelligence operations. In other words, the aim of speaker recognition is not to understand what was said, but who was saying it. This task is much easier to accomplish if the speaker willingly cooperates with the identification process, for example to gain access to a service he or she is entitled to. In that case, the speaker is usually enrolled in a voice authentication programme and has given consent to a recording of his/her voice of about ten second length to enable the software behind it to generate a voice template or voice print of about 1 KB size based on acoustics and speech analysis technologies (Myers 2004, 4, 7–9). This voice print then is entered into a data base and used for verification of the speaker's identity whenever he or she intends to make use of the particular service, for example accessing one's bank account. According to Myers (2004, 4), voice authentication has several advantages. First of all, unlike iris scans and fingerprints, speaking comes naturally and is thus widely accepted as a way of identification by the users of such systems. A second advantage mentioned by Myers is that voice authentication makes possible a remote identification via the phone – unlike in the case of fingerprint or iris-based recognition systems, the users does not need to be physically present to be identified. The authentication itself is both fast and reliable, and the authentication procedure as such basically is completed 'while we speak.' Seen from the perspective of user-friendliness, voice authentication thus is as hassle-free as it gets when it comes to identification and verification.

As regards real life applications, voice recognition and authentication technologies are already used by a number of high-street banks such as HSBC or First Direct in the UK, albeit not as a single security measure but as a second step in a two-step process – the first being the fingerprint scanner of one's smart phone. Apparently, the voice recognition technology used here is advanced enough not to be fooled by somebody mimicking somebody else's voice, and also works if the individual to be identified suffers from a cold since the voice print itself is not substantially affected by that. As a journalist pointed out in that regard, "[your] accent, cadence, pronunciation and physical attributes do not change even if you're a bit stuffed up" (Gibbs 2016). Also, the technology now seems to be capable of filtering out background noise – a disadvantage of voice recognition systems flagged up for earlier systems by Myers more than 15 years ago in 2004, who also argued that "[the] difference in telephones can be problematic [and that] enrolling with a home telephone and then trying to authenticate on a cell phone can be enough of a difference to cause a false rejection" (Myers 2004, 5). Having said that, even at the end of the second decade of the twenty-first century, Myers's caution that "[voice] biometrics is not the most secure of the biometrics technologies [which means] it is not appropriate to use them independently for authorization to systems that require high security" (ibid.) still makes eminent sense: a BBC journalist reportedly defeated the HSBC's system with the help of his non-identical twin mimicking the journalist's voice (Simmons 2017). Hence, it is fair to say that at the moment, voice recognition works best as part of a two-step verification and identification procedure, not so much as a standalone authentication feature.

It is however not only the way we speak, or the language we use that could be used for identification and verification purposes. Rather, it is the speed and rhythm of typing as well that, as a unique trait, also reveals to a sophisticated surveillance software who this person is that sits at a given keyboard in front of an identifiable computer or laptop or uses the virtual one of a mobile phone. Just listen to the rhythm of your typing, for example when you enter a password: yours is different from somebody else's rhythm, even if they would enter the same sequence of letters, numbers and special characters. Not that long ago, I have been treated to an advanced-stage demonstrator software that analyses one's writing pattern and rhythm and turns it into an algorithm to offer yet another layer of security: even if the user's password is compromised, the hacker who now is in possession of it still could not access the password-protected computer or bank account due to their different rhythm when it comes to typing the password. Again, the beauty of this solution is that unlike a password, we cannot normally 'forget' our way of typing – 'normally' here meaning that there are circumstances that might at least temporarily interfere with that, such as a sprained wrist or, maybe, just a new keyboard. Hence, this security feature is most helpful as one of several ways of identification to choose from – at least as long as we don't suffer from a severe laryngitis and a sprained wrist at the same time…

Since I already discussed hands in the context of typing and sprained wrists, it is time to add yet another biometric feature of our body that can be used, and actually already is being used, to identify individuals: this is the geometry of one's hands which also contains fairly unique features – although not as unique as fingerprints or iris patterns. Hand geometry technology usually is based on the recognition of palm and fingers structure: at an access control point, all one needs to do is to put one's hand on a hand geometry reader that compares one's hand with a saved pattern in order to decide whether access should be granted or denied. The geometry reader does that by taking a three-dimensional picture of the hand, measuring the shape and length of the fingers, and also the wrists in a process that lasts about five seconds (Bača et al. 2012, 78). This technology was developed in the 1970s and was first used in facilities at New York's Wall Street (Bača et al. 2012, 77). Due to its user-friendliness combined with ruggedness and reliability, it quickly found acceptance in many security environments, for example San Francisco International Airport to control access to its operation areas, or Israel's Ben Gurion Airport for the convenience of trusted travellers (Modi 2011, 103). As Boussadia adds, "[frequent] flyers and Israeli citizens can use the HandReader system for a small fee in conjunction with a credit card and a thorough background check to verify their identity, allowing security authorities to focus on unknown travellers who are considered to be higher risk, improving overall security levels and reducing waiting times" (Boussadia 2009, 7). A further advantage is that compared to other biometrics, especially fingerprints, hand recognition systems are usually found more acceptable since they have what Bača et al. (2012, 96) call an "appropriate privacy violation level", meaning that they are deemed to be less invasive than iris scans or fingerprints. The main disadvantages of hand recognition systems can be found in the somewhat limited uniqueness of hands: as Bača, Grd and Fotak (ibid.) point out, "[from] a hundred randomly chosen persons, at least two will have similar hand geometry" – which renders this technique comparatively easy to forge. Furthermore, and rather obviously, a hand injury will "have great impact on the recognition system" (ibid.). Nevertheless, when used in conjunction with other biometrics, hand recognition systems are quite useful.

Apart from measuring the hand's geometry and shape, two other technologies are used in that regard: palm technology that measures the "size, shape and flow of papillae and their distances", and hand or finger vein technology that uses "infrared light or some other technology for retrieving image [sic] of veins under the human skin" (Bača et al. 2012, 78–79). These variants are somewhat safer than hand recognition, face recognition, or even fingerprints: as Yang, Shi and Yang explain, "[in] contrast to conventional appearance-based biometric traits such as face, fingerprint and palmprint, finger-vein patterns are hidden beneath the human skin and unnoticeable without the help of some specific viewing or imaging devices" (Yang et al. 2012, 59). However, they hasten to add that "the captured finger-vein images are not always good in quality [which] certainly makes finger-vein feature representation unreliable" (ibid.). Again, this technology probably works best in conjunction with other biometrics – as a stand-alone, especially when it comes to high-security environments, this technology is too unreliable.

Another and maybe at first glance not so obvious biometric that can be used for identification and verification is one's gait, that is the way one moves and walks. And even though it is not an obvious biometric at first glance, people usually claim that they can identify people they know from a fair distance just by the way they walk and comport themselves. Hence, it is only logical that this very personal trait also is used to verify the identity of a known user. However, as Boyd and Little point out, one's gait is difficult to turn into a reliable pattern or gait print since it is hard to define to which extent a person's gait is unique – after all, "there are several factors […] that cause variations in gait, including footwear, terrain, fatigue, and injury" (Boyd and Little 2005, 19). Thus, Boyd and Little remain unconvinced regarding its usefulness for biometric purposes. It is not an impossible task, though. Again, the easiest way to turn one's gait into a biometric is via an enrolment programme aimed not necessarily at identifying an unknown user, but at verifying the identity of a known one. First, a gait print or gait signature is established by having the individual enrolling on the programme walking up and down for a certain amount of time so that the camera-based gait biometric system can create a pattern via an analysis of one's motions. This gait signature then is entered into a database and used for future verification – usually via cameras. Some other systems verify one's gait via floor sensors, or via wearable sensors (Mason 2016, 20). Although the latter approach may sound a bit outlandish at first, or at least very restrictive with regard to the number of potential users willing to don such sensors, the advent of smartwatches actually makes this approach rather feasible (Johnston and Weiss 2015). In the foreseeable future, we may thus see such systems being rolled out, probably also working in conjunction with other biometric security systems, at least as regards high security environments. Even though such systems have their limitations, they are very user-friendly and unobtrusive – just like voice recognition systems. Even more importantly for broader identification purposes of individuals not enrolled in any programmes, and again just like voice recognition, gait recognition systems do not require the active and voluntary participation of an individual about to be identified – rather, gait recognition can be used in conjunction with smart CCTV systems to home in and identify a suspicious person from among a crowd, as I will discuss later on. Hence, making gait recognition systems work has a certain appeal.

There is yet another biometric that could be used, and which is actually the most unique one of all, and that is one's DNA. Its uniqueness certainly is a great advantage when it comes to verifying and identifying individuals. On the other hand, it is clearly also the most obtrusive one of all since a cell sample has to be taken via a so-called 'buccal swab' that "involves wiping a small piece of filter paper or a cotton swab against the inside of the subject's cheek, in order to collect shed epithelial cells" (Hashiyada 2011). Currently, it is also by far the slowest with about at least five minutes extraction, testing and identifying time for the fastest system, while other systems might take up to 90 minutes to complete the procedure. In my opinion, it is less a question of speed than of intrusiveness that seems to rule out DNA testing for most identification and verification systems. After all, many individuals already show reservations when it comes to passing through a full-body scanner (this issue will be discussed below) – having to undergo a swab test would thus very probably be seen as a step too far. As such, DNA tests are more useful when it comes to forensic purposes, but not necessarily for general security procedures.

5.4 Towards Multimodal-Based Biometric Systems: Evaluation

At the end of this chapter, we cannot really avoid coming to the conclusion that biometrics alone are of limited value in the fight against terrorism. First of all, most of them are, to a certain extent, cumbersome, intrusive and/or more or less easy to defeat. And secondly, they all work best for verification and authorization purposes of known users whose biometric data have been collected via an enrolment process – either a voluntary one as in the case of voice recognition for banking, or a compulsory one in the case of facial and fingerprint recognition as used for biometric passports, in all probability soon to be followed by iris recognition. Only facial recognition, voice recognition and gait recognition are also useful when it comes to identifying unknown users since collecting these biometrics does not require the voluntary participation of individuals – smart CCTV systems (discussed below) can do that without anybody noticing. The problem with that however is (apart from human rights issues) one of available data that can be matched to the newly collected information of, say, an individual walking through the line of sight of such a smart CCTV system: sure, the gait and the face may well be unique to that person, but that hardly matters if there is no matching information available in a database. The same can be said for a voice print taken during a phone intercept as part of an investigation. Maybe that can be done in the future, and I will return to this issue below when I discuss data-mining, but at the moment, we are not there yet.

David Lyon shares this pessimistic but realistic feeling, bluntly arguing that "as tools for an anti-terrorist campaign they are flawed" (Lyon 2004, 310). As he further explains,

> [the] automated, algorithmic systems are poorly equipped, by and large, for the task of identifying or monitoring the actions of previously unknown terrorists. Moreover, to the extent that surveillance depends on information technologies, the easier it will be for persons who wish to evade detection to do so, just because human beings are more flexible and imaginative than technologies. Any technology can be outwitted, given time and ingenuity. (Lyon 2004, 310)

This is especially the case for the current standalone biometric security systems, none of which can be deemed to be absolutely secure: as we have seen, all of them have their pros and their cons. Al-Abed, Charrier and Rosenberger broke them down according to universality, uniqueness, permanency, collectability, acceptability and performance of detectors. They found that regarding performance, the interpretation and verification of biometric data by all current security systems is error-prone, with no system reaching a 100 per cent reliability. Regarding acceptability, they found that not all biometric security systems are accepted by the public, for example fingerprints ('fair'), iris recognition ('poor') or DNA recognition ('poor'). For the latter, they also evaluated the collectability of data as poor. Other biometric features such as gait, keystroke pattern, face or voice are obviously not permanent, while gait, face and hand geometry additionally lack in uniqueness as well (Al-Abed, Charrier and Rosenberger 2012, 151–152). All these issues obviously create security issues of various kinds – the most benign being false rejection rates (FFR) or false non-match rates (FNMR): here, individuals are not cleared by detection devices because the biometric they submit are not accepted. Worse however would be false acceptance rates (FAR) or false match rates (FMR) since in these cases, individuals manage to slip through security barriers who should have been rejected. Hence, Al-Abed, Charrier and Rosenberger's recommendation is to make use of multi-modal-based biometric systems – that is security systems that consist of several biometric scanning systems. To an extent, this approach is already applied by the most modern of biometric passports: here, face recognition now is combined with fingerprint recognition.

However, as Al-Abed, Charrier and Rosenberger also point out, such multimodal systems come with at least two drawbacks: "The first one is due to the cost that is, of course, increased as many sensors are necessary. The second one concerns the usability for users that have to make many operations to be authenticated" (Al-Abed et al. 2012, 163). While costs certainly are an issue, especially if and when there are no government subsidies forthcoming, the usability, or rather user friendliness, is an obstacle much more difficult to surmount: some readers such as iris recognition devices for example are not that fast, while also complicated to use. Combining two or more of them, or forcing individuals to go through several of them subsequently, will thus be time-consuming and frustrating. Given the propensity of users to get the not always clear instructions of even just one of such devices wrong, long queues will be the foreseeable result – which creates security issues in itself by offering a rather large soft target to terrorists. Hence, what is really needed is a 'one stop shop' that reliably identifies individuals without slowing them down. I will come back to this hoped-for solution and the chance to see it developed and deployed later on at the end of the discussion of scanners and sniffers.

There is yet another potential problem that Rathgeb and Busch flagged up: the crucial question of how to secure such data – especially data derived from a multi-modal biometrics system that checks and combines the results (or 'matching scores') of several biometric features (for example iris and fingerprint recognition) in order to arrive at a final access/non-access decision. As they warn, "[the] leakage of any kind of template information to unauthorized individuals constitutes serious security and privacy risks, e.g. permanent tracking of subjects without consent or reconstruction of original biometric traits (e.g. fingerprints or iris textures) might become a realistic threat" (Rathgeb and Busch 2012, 173). They also caution that although technologies have been developed to prevent the leakage of such data, "template protection schemes which provide provable security/privacy, and achieve practical recognition rates have remained elusive, even on small datasets" (Rathgeb and Busch 2012, 173–174). A safe and secure storage of sensitive data however is a problem not only for multi-modal biometrics systems but basically for all kinds of sensitive data, as we shall see in the next chapter.

At the time of writing, several high-tech firms are working on multi-modal biometrics systems that are meant to marry user-friendliness with data security. One of them is the British firm Object-Tech that is working with Dubai's Immigration and Visas Department (GDRFA) to develop a 'biometric border.' The 'biometric border' pilot project is described as follows: "For each passenger, the biometric border system generates a digital passport that contains the data currently stored in chips on existing e-passports, as well as includes biometric information such as fingerprints, iris scans and facial recognition" (Lee 2017). For the identification of a passenger (obviously only such passengers enrolled in a 'digital passport' programme), facial recognition is the most important element: after disembarking from the plane, and on the way to baggage claim, the passenger walks through a 'biometric border' walkway. While the passenger walks though this 'biometric tunnel', it takes a three-dimensional face scan via LIDAR technology and compares it with the data stored on the digital passport (McGoogan 2017). Since available information on this biometric border pilot project is still a bit vague, it is unclear whether this system also consists of a gate that only opens when the face scan data matches the one stored on the passport. Regarding data security, this digital passport is said to use 'blockchain' technology[1] which, at the time of writing, is deemed to be highly resistant to data manipulation.

Dubai is not the only forward-looking country regarding biometric borders: Australia also plans to develop a comparable system that "would let passengers 'literally walk out like at a domestic airport'" (McGoogan 2017). In Australia's case, the pilot project on this contactless border is expected to start in March 2019 (ibid.). In a sense, both projects would certainly be steps in the right direction – if they can be made work, that is: many ambitious projects in a similar direction foundered as soon as the system was tested outside of the laboratory in a 'real life' setting at a busy airport. Nevertheless, some optimism is probably in order here: sooner or later, the teething problems that such ambitious projects still face at the moment will be overcome. And after a successful testing of such projects at airports, and as long as the systems are not prohibitively expensive, we can also expect them to be used outside airports as well – basically everywhere where access control is relevant.

[1] A blockchain consists of blocks of data which are individually secured via cryptography and linked via 'hash points' containing an encrypted code (hash value) of the current block plus the hash code of the previous data block and a time stamp. See for example The Economist 2015.

Chapter 6
Prediction and Postdiction: Real-Time Data Mining and Data Analytics

Abstract In this chapter, I focus on data-mining and data analytics. It is obvious that without integrated databases set up for collecting, collating, managing and disseminating data derived from various identification, monitoring and surveillance systems, all efforts in this direction would be in vain – which is why we need to discuss such computerized databases first. But databases are just the beginning: nowadays, computers can tap into the World Wide Web to proactively search all kinds of social media including listen in on phone calls to detect suspicious behaviour, to profile individual terrorists or suspects. This process is known as data mining and acquisition, and similar to what many private companies such as Google or Facebook routinely do ways as well. I argue that currently, the 'holy grail' of data-mining and acquisition is to be able to do so in real time, while 'it' happens. With a discussion of this current cutting edge of research and the implications for our civil liberties, I conclude this chapter.

Keywords Databases · 'Inspector computer' · Data-Mining · Digital footprints · Big data · Internet of things · Postdiction · Prediction · Profiling · Data-Mining · Civil liberties

What we learned in the previous chapter is that currently, biometrics are of limited use only when it comes to protecting ourselves, and our societies and cities, from terrorism: apart from the technical flaws that do not need to be repeated here, most biometrics work best in the context of verifying known users who either possess biometric passports or IDs or are enrolled in a biometrics-based security programme. Identifying unknown individuals remains difficult – although face recognition, gait recognition and voice recognition as methods that do not require the voluntary participation of an individual are promising in this regard. In a subsequent chapter, I will discuss smart CCTVs that could, for example, pick up and zoom in on a suspected terrorist after having recognized him or her because of the terrorist's unique gait or facial features. However, without integrated databases set up and optimised for collecting, collating, managing and disseminating data derived from various identification, monitoring and surveillance systems, all efforts in this direction would be in vain – which is why we need to discuss such computerized

P. Lehr, *Counter-Terrorism Technologies*, Advanced Sciences and Technologies for Security Applications, https://doi.org/10.1007/978-3-319-90924-0_6

databases as well. But databases are just the beginning: for the pioneers from the ranks of the police who started to set up and use such databases in the early 1970s, it was clear from the outset that these new computer systems could do much more than simply collect and store data: rather, they could also sift through vast amount of data to establish patterns not obvious to even the most seasoned police officer, and could thus be used not only to combat crime, but also to predict and prevent at least some of the manifestations of crime. Hence, for these far-sighted pioneers, it was obvious from the start that computers would be invaluable tools for political manifestations of crime, that is terrorism and political violence, if their full potential was realized.

More than four decades later, these early days of computerized policing look rather quaint when one examines the pictures of banks of computers fed by magnetic tapes. Nowadays, computers can tap into the World Wide Web to proactively search all kinds of social media or listen in on phone calls to detect suspicious behaviour, to profile individual terrorists or suspects, and to carry out a network analysis: who is talking with whom, where, and about what? This process is known as data mining and acquisition, and similar to what many private companies such as Google or Facebook routinely do in various ways: automatically collect and collate vast amounts of data about individuals in order to bombard them with personalized advertisement. Of course, the powers of intelligence services go further than those of private sector ventures, and the purpose is a different one – but the underlying mechanisms are the same. Currently, the 'holy grail' of data mining and acquisition is to be able to do so in real time, while 'it' happens. With a discussion of this current cutting edge of research that brings together the forces of law enforcement, intelligence services, and science, I shall conclude this chapter.

6.1 'Inspector Computer': Computerized Databases as Invaluable Tools for the Police

The realisation that databases would render the fight against terrorism, especially international terrorism, somewhat easier is not exactly a new one. First efforts to use databases of the 'pen-and-paper' kind were made more than one hundred years ago during the first global war on terrorism – that is, the global war on anarchist terrorism at the end of the 19th and the start of the twentieth century. Concerted European efforts to combat this first wave of modern terrorism were planned and discussed at the already mentioned International Conference of Rome in November and December 1989. As Deflem notes, "the protocol from the Rome Conference included provisions to encourage participating governments to have police keep watch over anarchists, to establish in every participating country a specialized surveillance agency to achieve this goal, and to organize a system of information exchange among these national agencies. […] In addition, the conference also approved a provision to extradite persons who had attempted to kill or kidnap a sovereign or head of state" (Deflem 2005, 278). Under the impression of the assassination of President McKinley in 1901, the USA led by new president Theodore Roosevelt joined the so far solely European effort. The St Petersburg Protocol of 1904 as the result of renewed negotiations included the proviso to establish central national agencies responsible for an institutionalized gathering and dissemination of information on anarchist activities (Jensen 2001, 21) – an exercise that depended on the collection of names and all snippets of information related to them. However, all that came to naught: due to deep-seated distrust to Imperial Germany and Russia, the USA refrained from joining the anti-terrorist alliance – which was made obsolete by the outbreak of the First World War in any case.

As concerns computerized databases, the first European one purposely created as a tool in combatting terrorism was set up in the early 1970s by the German *Bundeskriminalamt* (BKA, Federal Criminal Police Office) under the far-sighted directorship of Horst Herold (Chief Commissioner from 1 September 1971 to 31 March 1981). Herold could draw on the experiences and valuable insights he had gained during his time with the police of Nuremberg from 1965 to 1971 – one of the first police forces in (West) Germany to experiment with electronic data processing based on computers. Finding a new approach to data collection and collation was of utmost importance from the 1960 onwards when the police (not only in Germany) was faced with ever-rising rates of crime and a corresponding never-ending flood of reports and notifications that could not possibly be sifted through by over-worked police officers. This lead to the pessimistic impression that crime could no longer be solved, left alone prevented, but only managed (Seiderer 2004, 320–321). In these years, Herold oversaw the gradual introduction of a computerized central database including a fingerprint classification program ('Fingerabdruck-Klassifizierungsprogramm' in German). Due to the immense costs and, even more importantly, the sheer size of the early computers of the 1960s and 1970s, the police of Nuremberg had to share the resources of the municipal computer centre with other authorities, having access only four to ten hours per week on average to the 3rd generation Siemens 4004/35 system (Seiderer 2004, 322). In 1969, the police database of Nuremberg consisted of three 'Grundbänder' (basic magnetic tapes): a 'Fallband' (case tape) that could accommodate 150,000 acts of crime, a 'Daktyloskopieband' (fingerprint tape) with a capacity of 1.5 million individual fingerprints of 150,000 persons, and a 'Straftäterband' (criminals tape) of up to 95,000 persons (Seiderer 2004, 325).

Unlike other senior police officers who essentially regarded computers as nothing else but more efficient tools for otherwise unaltered procedures of registering, searching and comparing data, Herold knew from the very beginning that computers could do far more than that, and that their main use would lay in their ability to link all saved data, which was, in theory, limitless. In a nutshell, computers were valuable tools not only for a quantitative *search*, but also for qualitative *research* (Seiderer 2004, 327). He also was convinced that modern information technology would allow the police to become pro-active as well to tackle the root causes of various crimes. In his view, the German police should thus use this unique opportunity offered by computers to explore the criminals, their personal traits and their social environments or milieus, to analyse the reasons behind his actions, and thus to put the task of crime prevention on the same level as crime prosecution (Spiegel 1971).

As the new Chief Commissioner of the BKA, Herold lost no time to turn this previously rather under-used institution into Germany's leading data acquisition and procession centre for pro-active and preventative policing. In 1979, a stocktaking of the on-going computerization found that the system consisted of 37 data files with 4.7 million individual names plus the names of about 3100 organizations, a fingerprints section with datasets of 2.1 million individuals, and a photograph section containing the pictures of no less than 1.9 million individuals (Aust 2008, 141). Another file of more than 3500 people that contained their names, short descriptions, and a list of all available data useful for their identification (photographs, fingerprints, samples of their hand writing) was held in the new Personal Identification Centre that had been established in the wake of the assassination of Siegfried Buback, Germany's Attorney General, on 7 April 1977 (ibid.). In September of the same year, Herold's sophisticated computerised *Personen, Institutionen, Objekte, Sachen* ('Persons, Institutions, Objects, Items') or PIOS police investigation also was used to try to locate Hanns-Martin Schleyer who had been taken hostage by the Red Army Faction (RAF). Success was only thwarted by the unfortunate fact that the most important hint on Schleyer's whereabouts got lost within the police bureaucracy way before it could have been fed into the PIOS system (Prantl 2017). This demonstrates that even in a time of high-tech, Murphy's Law still applies – whatever can go wrong still will go wrong. Even so, the BKA's computer system "simultaneously fulfilled two of a detective's dreams: the collection of as much information as possible, and the ability to fit the independent components together in the minimum of time" (Aust 2008, 141).

Herold's pioneering efforts earned him glowing praise and many colourful epithets, such as 'Kommissar Computer' (inspector computer), 'police futurologist' or 'pioneer of electronic criminology'. It also earned him his fair share of criticism. For example, renowned German author Hans Magnus Enzensberger (1979) criticized that the underlying motto of this new information technology in the hands of the police seemed to be 'to collect as much as possible, and to delete as little as possible', while intruding deep into citizens' privacy. This concern was shared by many of Herold's contemporaries, who also feared that the tendency to collect ever more data going hand in glove with a steady widening of key concepts such as terrorism to also include mere sympathisers could result in a gradual criminalisation of ever more otherwise loyal citizens via a 'preventative stigmatisation' as potential 'enemies of the constitution' (Kreissl 1981, 137). Not surprisingly, some critical voices quickly drew comparisons with Germany's dark past. Enzensberger as one of Herold's fiercest opponent lost no times to connect the dots in this regard. His core arguments against the use, or rather, abuse, of information technology by the police was as foresighted as Herold's vision – albeit diametrically opposed to it:

> The classic form of repression never enjoyed such a broad support. A police force that shows itself openly and brutally on the streets always polarizes; it mobilizes millions of people against itself and creates deep-seated, long-lasting conflicts. Its logic is that of a

> latent civil war. The new, 'scientific' methods of social control however aim at integration; they are too clinical and not bloody enough to spark mass feelings of hatred and solidarity. The megabits of information that enter a computer every hour, unnoticed and silently, do not provoke riots. After all, they also guarantee that pensions are paid on time, and the health bureaucracy will repay the money spent for sleeping pills. (Enzensberger 1979) (translation from German mine)

About four decades later, Herold's trailblazing databases appear rather puny when compared to the vast relational databases of our times. For example, in 2014, the FBI's *Next Generation Identification* (NGI) programme as the "world's largest biometric database" already "held 73 million fingerprints, 5.7 million palm prints, 8.1 million mug shots and 8.500 iris scans" (Chakya 2014). The NGI programme shows that Herold proved to be rather prescient when he opined in the late 1960s that computers would allow for a limitless storage of data as well as a limitless procession of such data, and that humans would cause the only limitations, not technology (Seiderer 2004, 332). This limitless storage and procession of data which he envisaged is now possible – after all, computer technology has come quite a long way since the 1970s. The emergence of such all-encompassing relational databases that do not only contain millions of names but hundreds of millions of them is not necessarily due to malign intentions of modern states hell-bent to making Orwell's vision expressed in his novel *1984* come true. Of course, the threat of international terrorism à la Al Qaeda and ISIS greatly helped in that respect, even though the threat may well be a bit over-blown as John Mueller (2006) argued. And of course, sweeping advances in computer technology also greatly facilitated the collection of ever more data on a scale that Herold could only dream of. However, and here lies the crux of the issue, these sweeping advances in computer technology also changed us, and the way we lead our lives – not only 'we' as innocent citizens, but also 'they', that is, in our context, the terrorists.

6.2 Digital Footprints, Big Data and Data-Mining: The Current Situation

Most of us nowadays tend to pay smaller amounts not with hard cash but by virtually (literally, even) 'waving our money good-bye' with the use of a contactless credit card. We also use our advantage cards, club cards, nectar cards etc. issued by department stores in order to get discounts or points. We use key cards to get into buildings. We use our oyster cards or equivalents to use busses, subways and railways. If we travel, we identify ourselves with passports likely to include biometric data, and we present our frequent flyer cards. We check news online, watch videos on YouTube, we update our Facebook or Instagram accounts or put a message out on Twitter, and we chat with others on Messenger, WhatsApp, Viber or Line, routinely (and sometimes obsessively) using our smart phones and the multiple apps on them. All these activities have one thing in common: they leave traces in the shape of electronic footprints. In a sense, we have become 'digital persons', as Daniel Solove argues, with the consequence that "[every] day, rivulets of information stream into electric brains to be sifted, sorted, rearranged, and combined in hundreds of different ways" (Solove 2004, 1), exactly as predicted by Horst Herold. Hence, by spreading electronic information about ourselves wherever we go, we enable interested parties to create what Solove calls 'electronic dossiers', that is electronic "collection[s] of detailed data about an individual" (ibid.).

'Interested parties' means that it is not only the state who can create such 'electronic dossiers' about us: private companies do so as well. Again, their interest in our data is not necessarily a malign one. Our banks and credit card companies have an understandable interest in keeping track of our spending, for example. And those companies whose advantage cards we have also need to keep track of our purchases to update our accounts. They can however do much more with our data – for example analysing our shopping habits to send us tailor-made offers. Or, and here it gets a bit more sinister, at least a bit annoying for most of us, they can sell the data generated about us to third parties so that they can bombard us with equally tailor-made advertisement as soon as we go online. Our Google search histories, and our Facebook timelines are telling examples: what we get as advertisement depends on the searches we made, on the 'likes' we awarded, and the personal interests (films, videos, music, sports, hobbies, etc.), as well as on our age, our gender and sexual preferences, in case we divulged this information. In the words of Iain Macwirther, writing in the aftermath of the Cambridge Analytica scandal (discussed in the third part):

> We know that Facebook holds 98 data points about each of us, including political outlook, purchasing history, hobbies, religion, location, age, gender etc. Google also collects data from billions of our searches on everything from herpes to Hollywood. In the era of Big Data, this information can be cross-referenced with other databases on credit rating, criminal history (even if you've never been prosecuted) and online health records to generate an astonishingly detailed picture of every single one of us. (Macwirther 2018)

All this information we have volunteered without anybody forcing us to do so, turning ourselves into ever more digital persons, most of the time without even thinking about it. But as Macwirther indicated, even that is only the tip of an iceberg of data available about us: via partnerships with other companies, for example credit reference agency Experian, Facebook also is likely to have a fair idea about our income, the credit cards we have and our credit lines. If we would ask for the data Facebook has collected about us, we would get a massive pdf file that includes "records of every person who had ever poked [us], all the IP addresses of machines [we] had used to access the site (as well as which other Facebook users had logged in on that machine), a full history of messages and chats and even [our] 'last location'" (Solon 2012; also see Lee 2014).

Again, nobody is forced to join Facebook, Instagram, Snapchat, WhatsApp or other manifestations of social media, and leaving such sites is entirely possible: there are many websites including YouTube clips guiding us step by step through the not-so-straightforward process of leaving Facebook for good, for example, and there are many more websites that teach us how to surf the Internet unobserved, and what kind of Virtual Private Network (VPN) or anonymizer software to use for that purpose. However, hiding our traces does not necessarily mean that these traces are gone completely: it just means that tracking us is made a bit more difficult – that's basically all. This is quite an important issue, flagged up for example by IT specialist Karsten Nohl, interviewed by Xifang Yang:

> 'Trying to move entirely invisibly through our digital presence does not work anyway', opines [IT specialist] Karsten Nohl. Not by using old Nokia handsets, not by VPN programs, not by the anonymizer network TOR. 'Even if you would do all that, you would still use the same web pages, contact the same people, in the same language, with the same speed of writing. If you would operate a drugs-trafficking network on the Dark Net, maybe then you could stay anonymous. But not if you do all those things in the Net that characterize you as a person.' (Yang 2016, translation from German mine)

Indeed, the way we move through the digital mirror image of our 'real' lives establishes a pattern via a series of digital footprints – or, in another figure of speech, digital finger prints. After all, as Nohl mentioned, it is not only the way we speak and the language we use that gives us away but also the way we type when we, for example, enter a password. As an additional security feature, an algorithm based on our unique way of typing can be a very useful additional security feature for verification purposes, as I argued above. This also implies however that the same pattern or algorithm can be used to identify the individual just entering data via some keyboard, even a virtual one, without the individual being aware of this. After all, this unique pattern stays the same even if we use an anonymizer software such as TOR that is meant to cloak our online activities.

A new technological trend is bound to further increase the digital foot prints we are leaving when we move through the digital or cyber world – and that is the vaunted 'Internet of Things.' In the introduction, I already mentioned the 'agency of things' when I discussed how technology changes our behaviour, usually without us noticing this. The 'Internet of Things' simply refers to all the high-tech gadgets now available that are meant to make our daily lives easier by improving overall accuracy and efficiency in conjunction with a reduced requirement for us directly interacting with them. For example, so-called 'smart meters' in the shape of sensors-equipped thermostats could monitor local weather and adjust its temperature settings without our interference. They could also increase the temperature, remotely kicked into action by us via a software application or 'app' for short, while smart kettles could get the water boiling at the same time so that we can get our cup of tea just when we arrive. Smart refrigerators can detect that we are out of fresh milk, or that the milk we have expired, and automatically order new milk over the Internet. And devices like Amazon's 'intelligent personal assistant' Alexa allow us to control many of these smart appliances in our home via your voice – as does Apple's Siri for the apps on iPhones, and Google Now and Cortana do for Android phones.

If we are driving in a car to work, our car's computer (connected to the Internet) will check the traffic services to alert us of traffic jams, while also suggesting ways around them. If we are homeward-bound, it will also alert the gadgets at home of our arrival time or of possible delays – which is why the kettle will just have boiled whether we are earlier or later than expected. When being 'out and about', the pictures we snap will be uploaded to the 'Cloud' so that all our linked devices such as home computers, laptops and tablets have access to them, while pedometers and heartrate monitors as part of our smart watches or smart phones can measure our activities and upload these information into our health account at our GP's practice. And on the basis of how many calories we have burned (or not), personalized advertisement in our shopping malls of choice will recommend certain food items while ignoring others. Of course, since our wearable devices or smart phones know where we are via Global Positioning System or GPS tracking, it might also suggest a restaurant nearby, based on our preferences, and whether we are in a car or on foot. Regarding shopping malls, the already mentioned RFID chips offer the convenience of an instant hassle-free checkout: the probably fully automated checkout already knows how much we are due to pay and will automatically charge our debit or credit card of choice while we casually walk to the exit (Blunden 2018) – at least, that is the plan for the near future since at the moment, only few shops have taken the necessary steps to enable this novel way of shopping.

Regarding safety and security, modern car computers are already able to automatically dial emergency call centres in case of a major accident, also giving the car's precise location. The same service is on offer for the elderly and infirm living alone via wearables, for example as a pendant necklace, or as a bracelet: all that is required in an emergency is hitting a button, which triggers an automated distress call combined with the precise location of the wearer. Nowadays, such emergency systems are also being integrated in the backpacks or school uniforms of children, also allowing them to alert first responders via a simple push of a distress button. Less welcome for the average user (probably the majority of us) might be yet another set of wearables that also allow for the tracking individuals: smart ID cards fitted with RFID chips issued to access one's work place. One the positive side, these smart cards are very handy since they open doors or switch on lights and air conditioning systems automatically, usually without the need to get them out of the wallet. In the case of an emergency, these smart ID cards will also alert first responders of one's presence and of one's precise location. Unfortunately, and this is the negative side, all this information is, in theory, also available for our employer who will know if and when we leave the premises for a sneaky cigarette or how often we use the rest rooms (Blunden 2018). In that particular case, it is quite obvious that there is a flipside to the convenience offered by the Internet of Things, courtesy of the cheapness of computer chips. In a sense, this is one step further towards a future as depicted in the Hollywood blockbuster *Minority Report* in all its positive and negative aspects.

The rather recent emergence of this 'Internet of Things' is made possible mainly because of the availability of cheap processing power. In the words of McEwen and Cassimally, this means that we "can afford to put processors in places that [we] couldn't before – not just a business workstation or a home PC but also a telephone, an electricity meter, a bedside lamp, or a teddy bear. We can make objects intelligent. We can make them think and speak" (McEwen and Cassimally 2014, 1). And via Wi-Fi, Bluetooth or 3G/4G, we can also connect them to the Internet and let them stay connected 24/7 – also made possible by the drastic decrease of costs related to Internet connectivity when compared to the early days (the 1990s) when people had to dial in via a modem. Hence, as McEwen and Cassimally put it, the Internet of Things is an equation of physical objects plus controllers, sensors and actuators plus the Internet (McEwen and Cassimally 2014, 11). In our context of urban societies, Williams and Selle emphasize the importance of the Internet of Things by linking it to the novel concept of 'smart cities', defined as "cities that are not only environmentally friendly, but also integrate and exploit high technology to control power, flows, and the delivery of services and goods" (Williams and Selle 2016, 11). Obviously, as Williams and Selle further elaborate, these smart cities are dependent on the Internet of Things, "which is also likely to be largely urban in character, or at the very least, most powerful and pervasive where there are large concentrations of people, homes, and businesses" (ibid.). They however also point at yet another trait of the 'Internet of Things' in conjunction with 'smart cities' – which is its Janus-faced character of intrusion and surveillance on the one hand, and safety and security on the other:

> There is, of course, the possibility that even smart cities might have their own dystopian underside where dense connectivity and transparency become intrusion, surveillance, and oppression. Yet, it is also clear that smart cities are most likely to be efficient, well-functioning, and, for the most part, able to provide high levels of security and safety to their citizens. (Williams and Selle 2016, 11)

Whether we are aware of it or not, the 'Internet of Things' is bound to greatly expand our digital foot prints, resulting in a flood of ever more information about us, our lives and our hobbies and habits. These huge amounts of information generated by our modern ways of digital communication via social media, smart phones, wearables (e.g., smart watches), smart homes and meters, credit card transactions are also known as *Big Data*. The main difference of Big Data to 'conventional data' is not necessarily its sheer size, but its short-lived, weakly structured/unstructured and ephemeral nature of most of it, which renders it too difficult to enter in normal databases. Hence, Big Data is understood as information marked by the three dimensions of *volume* (the amount of data collected), *velocity* (to speed of data generation and data transfer) and *variety* (the sheer width of data types and data sources) (Christl 2014, 12). Making sense out of that kind of data, for example to put tailor-made advertisement deemed to be relevant for us on our personal Facebook page, predicting our next location depending on the recent and current location reported by our smart phones in order to suggest a bar or restaurant, or even judging our current emotion from the pattern and speed of our typing (Christl 2014, 21–22) is an automated process known either as *data analytics* or *data mining,* based on computer algorithms oftentimes written in programming languages such as SQL, Python or R and sold as ready-to use-software packages such as Tableau, JMP, SAS or Windows Analytics.

6.3 Anti-Terrorism Data-Mining and Acquisition: The (Current) Cutting Edge of Policing

If Facebook, Google and other private companies can automatically and fairly reliably mine or analyse Big Data via the use of clever computer algorithms coupled with the Internet and the Internet of Things, then it is more than obvious that the state, via its many agencies, can do the same – actually, much more of the same. Although, as already mentioned, state authorities such as police and intelligence forces realized the importance of computers and information technologies long before the advent of Al Qaeda, the events of 9/11 certainly accelerated the process. Most importantly, it accelerated the growth of a field of study known as Intelligence and Security Informatics (ISI) – a comparatively new science that, like terrorism studies, rapidly mushroomed into a major industry as a result of 9/11. The newly created Department of Homeland Security (DHS) was instrumental in that regard, defining six areas in which ISI should play a leading role: intelligence and warning, border and transportation security, domestic counter-terrorism, the protection of critical infrastructure and key assets, the defence against catastrophic terrorism using biological, chemical, radiological or nuclear weapons, and emergency responses and preparedness.

Regarding intelligence and warning, the underlying assumption was (and still is) that since terrorist attacks are premeditated and not random, it should be possible to detect them in the planning stage. After all, targets need to be selected and reconnoitred, training has to be conducted, and weapons need to be acquired or IEDs to be built – all of which leaves digital footprints of one kind or another since terrorists are also prone to use the same services that everyone else uses, including the Internet, and the Internet of Things. Of course, it is possible that an attack is planned, prepared and carried out by a so-called 'lone wolf' who acts completely alone and without leaving any traces – Anders Behring Breivik would be an example for that. But most of the time, terrorism is a group-based activity in which several group members carry out specific tasks, which makes them vulnerable to detection. As Chen explains, "[terrorism], like gangs and narcotics trafficking, is regarded as a type of organized crime in which multiple offenders cooperate to carry out offenses. Information technologies that help find cooperative relationships between criminals and their interactive patterns would also be helpful for analysing terrorism" (Chen 2006, 4). And in case some of the participants in a plot have to travel to reach the location where their intended target can be found, they are bound to leave additional digital footprints when they cross borders, especially 'smart' ones featuring e-gates and the like, as discussed above. But even in the case of lone wolves such as Breivik, more often than not there are potentially detectable traces left during the preparation of the attack, for example Google searches, or the act of buying certain ingredients required for home-made explosives via the Internet, or maybe just by renting a van or a car using a credit card.

Our ever-increasing digital footprints makes it comparatively easy for state agencies to set up immense databases which are of eminent importance for intelligence and warning as envisaged by the Department of Homeland Security (we shall leave their impact on civil liberties aside for the moment). For example, in the case of the USA in the aftermath of 9/11 and the passing of the controversial *Uniting and Strengthening America by Providing Appropriate Tools Required to Intercept and Obstruct Terrorism Act* (USA Patriot Act for short), the US Department of Defense developed a computer program initially called *Total Information Awareness* (TIA) before its name was changed to the somewhat less contentious and 'Big Brother' sounding *Terrorism Information Awareness* program – still abbreviated TIA, as Fienberg (2008, 200) points out. Before it was finally stopped by Congress, this program basically collected personal information held by databases of private companies to turn them into massive electronic dossiers of individuals. These dossiers where then sifted through to detect individuals who were "likely to be engaged in criminal activity" (Solove 2004, 5–6).

Again, we find Herold's 'Inspector Computer' at work, albeit at a much grander scale. However, as Solove emphasizes, these massive dossiers remain meaningless if they are not "sifted, sorted, rearranged, and combined in hundreds of different ways" (Solove 2004, 1). For example, the act of renting a van or a car in itself may look innocuous unless it is linked to a credit card that has been flagged up before, or to an individual under suspicion for some reasons – maybe just because the individual chatted with the wrong kind of people before renting the vehicle. While seen in isolation, all these bits and pieces of information might look harmless enough, in combination they could be the tell-tale signature of an impending car bomb attack. Basically, and in the words of former Department of Homeland Security Secretary Michael Chertoff (2006), the gist of such data-mining operations is "to thwart a terrorist who has not yet been identified." This 'thwarting' can be done by immediate police intervention if the risk of an attack is deemed to be imminent, or by putting a suspect under surveillance and possibly also on a terrorist watch list. An example for such as system is the US Department of Homeland Security's *Automated Targeting System* (ATS). As Amoore and de Goede explain,

> ATS analyses a variety of passenger data – including address, financial records, 'no show' history, how tickets were purchased, motor vehicle records, past one-way travel and seating preference – in order to assign a risk score to individual travellers. The score is used to determine whether passengers or border crossers are placed on a 'selectee list' for further attention, stopped and questioned at the border or, indeed, denied entry. The risk assessment calculation in ATS is classified, and the results can be kept on file for up to forty years. (Amoore and de Goede 2008, 5)

As already mentioned, the process of 'sifting through' is better known as 'data analytics' or 'data-mining'. According to Feinstein and Kaplan, this was, at least initially, not done – rather, the focus was on the mere collection (or "overcollection", as they call it) of ever bigger amounts of data on the one hand, mismatched by a lack of human resources to analyse them in any meaningful way on the other (Feinstein and Kaplan 2011, 281–282), basically looking for the proverbial 'needle in a haystack' without knowing whether there would actually be a needle in the first place. To achieve the proper balance between data collection and data analysis without having to recruit thousands of junior and senior analysts, and to establish meaningful patterns about potential persons of interest, state authorities also quickly turned to using an automated process of 'data-mining' or 'data analytics' that involves human operators only at a very late stage – that is, when something of relevance has been found. In the case of the USA, this was achieved by the successor of both TIAs named *Multistate Anti-Terrorism Information Exchange* or MATRIX, in use between 2005 and 2006 before it was terminated due to privacy concerns (Fienberg 2008, 200; Department for Homeland Security 2006).

When talking about data-mining or data analytics, we basically mean a process also known as 'Knowledge Discovery from Databases' (KDD for short) that turns raw data into meaningful information (Gandy 2002, 3). As Chen explains, "[knowledge] discovery refers to non-trivial extraction of implicit, previously unknown, and potentially useful knowledge from data" (Chen 2006, 17). In the context of countering terrorism, the underlying assumption is that "prospective datamining could be used to find the 'signature' of terrorist cells embedded in larger networks" (Fienberg 2008, 197). There are several techniques of data mining which are relevant in the context of ISI, and, thus, also in our context of defending ourselves against the threat of terrorism. The first technique is *text mining*, a potentially multilingual technique which is "critical for extracting key entities (people, places, narcotics, weapons, time, etc.) and their relationships presented in voluminous police incident reports, intelligence reports, open source news clips, etc." (Chen 2006, 19). Xiaohu Hu et al. offer an interesting example on that in the case of text-mining open-source biomedical literature for the purpose of identifying potential viruses or bacteria that could be weaponised for bio-terrorism (Hu et al. 2008, 385–406). The second relevant data mining technique is *spatial and temporal mining and visualization,* in order to support "geographic information systems (GIS) and temporal analysis of criminal and terrorist events" (ibid.). Here, Chen criticizes that while "[most] crime analysts are well trained in GIS-based crime mapping tools, […] automated spatial and temporal pattern mining techniques (e.g., hotspot analysis) have not been adopted widely in intelligence and security applications" (ibid.). Finally, the third relevant data-mining technique is *criminal (social) network analysis*, important when it comes to detect who is interacting with whom. In Chen's words, "[often] referred to as 'dark networks,' these organizations exhibit unique structures, communication channels, and resilience to attack and disruption" (Chen 2006, 19–20). Extensive social network analyses have for example been conducted in the aftermath of the 9/11 attacks, and after the London 7/7 bombings, to discover with who else the immediate perpetrators were in contact prior to the attack, in order to unravel the whole terrorist network behind these incidents. Currently, many more data-mining architectures and automated processes are employed or under development to deal with ever vaster amounts of data which cannot possibly be discussed here without turning this book into a technical compendium. For interested readers, the open source publication *Emerging Technology and Architecture for Big-data Analytics* of Chattopadhyay Anupam, Chip Hong Chang and Hao Yu (2017) offers an excellent overview.

Above, I already mentioned MATRIX, which, although now defunct, is of eminent interest in the context of data mining. Fienberg (2008, 201) draws our attention on Relyea and Seifert's report for the Congressional Research Service that amongst other background information on that programme also describes its analytical core (i.e., a set of algorithms) that actually carried out the data mining required. Relyea and Seifert note that the analytical core of MATRIX was a software tool named *Factual Analysis Criminal Threat Solution* (FACTS for short). By automatically sifting through dynamically combined records retrieved from thousands of unconnected data bases containing "over 3.9 billion public records", FACTS enabled authorized users to assemble partial information into a complete picture of persons of interest (Relyea and Seifert 2005, 12–13). Of additional interest is the fact that MATRIX was developed not by one of the US intelligence services but by a company named *Seisint* – one of many private companies developing data mining tools and software, used not only by state authorities but by private sector enterprises as well. But I will deal with the implications of this below. For the time being, let us return to Fienberg and comment on the – now rather obvious – goal of data mining: "the identification of terrorists or criminals in a way that would not be possible from the individual data bases" (Fienberg 2008, 202) as these only possess partial information that in itself is insufficient to paint a coherent picture of potential malicious activities, and the individuals behind them. Fienberg further explains that there are two different aspects to this goal: "(1) identification of known terrorists which is a form of retro- or postdiction, and (2) identification of potential future terrorists and profiling, which involves prediction" (Fienberg 2008, 202). The latter one usually draws the ire of civil liberties organisations due to its potential infringements on citizens' right to privacy, but for the purpose of nipping terrorist attacks in the bud, it is by far the more important goal, and the most promising one as well. Not surprisingly, prediction is what two of the most talked-about (in recent times) data-mining programmes run by the USA's NSA and the UK's GCHQ respectively aim to achieve: *Prism* and *Tempora.*

NSA's *Prism* program, launched in 2007 during the presidency of George W Bush, was disclosed to the public as part of the security leaks courtesy of whistleblower and ex-NSA contractor Edward Snowden. Basically, the program functions as a large data-mining operation, just like those conducted by private enterprises whose role I discussed above. The difference is that NSA's Special Source Operations Group, which is responsible for Prism, has "far more privileged access" (Gellman and Poitras 2013). For example, Prism is said to tap "directly into the central servers of nine leading U.S. Internet companies, extracting audio and video chats, photographs, e-mails, documents, and connection logs that enable analysts to track foreign targets" (ibid.). The servers mentioned are those of Microsoft, Yahoo, Google, Facebook, PalTalk, AOL, Skype, YouTube and Apple (ibid.). As Greenwald and MacAskill assert, "with this program, the NSA is able to reach directly into the servers of the participating companies and obtain both stored communications as well as perform real-time collection on targeted users" (Greenwald and MacAskill 2013). Prism is not necessarily a free-for-all, though: its use is governed and controlled by the US Foreign Intelligence Surveillance Court (FISA Court), and safeguards are in place to minimise data collection of US citizens. Nevertheless, for obvious reasons (potential civil liberties infringements, for example) NSA would have very much preferred to keep the program secret. As the Director of National Intelligence, James Clapper, pointed out, "information collected under this program is among the most important and valuable foreign intelligence information we collect, and is used to protect our nation from a wide variety of threats. The unauthorized disclosure of information about this important and entirely legal program is reprehensible and risks important protections for the security of Americans" (Gellman and Poitras 2013). Seen strictly from an anti-terrorism perspective, and in the context of high-tech as a silver bullet in the fight against terrorism, Clapper is not wrong: the mere knowledge that such a program exists plus the information available now allows terrorists, especially high-end ones who also make use of the most advanced technological means they can get access to, to develop plans to circumvent its unique strengths and data-mining strategies.

It is however not only the USA's NSA that uses huge data-mining operations akin to immensely magnified versions of the early-day dragnets implemented by the German BKA in the 1970s: at the same time when Prism was leaked, documents related to Snowden revealed that one of the three UK's intelligence services, the *General Communications Headquarters* (GCHQ), also had developed operations similar to Prism – not necessarily comparable in size, but obviously also not that far behind. As an article in the *Guardian* puts it, "[the] sheer scale of the agency's ambition is reflected in the titles of its two principle components: Mastering the Internet and Global Telecoms Exploitation" (MacAskill et al. 2013). GCHQ's ambition goes much further than conducting conventional data-mining that involves sifting through thousands of unrelated databases. Rather, GCHQ has developed the capability "to tap into and store huge volumes of data drawn from fibre-optic cables for up to 30 days so that they can be sifted and analysed" – an operation code-named *Tempora* (MacAskill et al. 2013). The scale of the operations, the results of which are shared with the NSA and, probably on a 'need to know' basis, with other intelligence services that are part of the *Five Eyes* intelligence sharing alliance consisting of the UK, the USA, Canada, Australia and New Zealand, can be gleaned from yet another quote from the Guardian article: in 2012, GCHQ was allegedly "handling 600m 'telephone events' each day, had tapped more than 200 fibre-optic cables, and was able to process data from at least 46 of them at a time" – which is why the authors of the Guardian article call the GCHQ an "intelligence superpower" (MacAskill et al. 2013).

Compared to NSA and GCHQ, other intelligence services' data-mining efforts appear puny in scale and ambition. For example, Germany's foreign intelligence service, the *Bundesnachrichtendienst* (BND, Federal Intelligence Service), is comparatively small, relatively underfunded and largely shunned by experienced IT specialists as a possible employer (Mascolo et al. 2017). Hence, it has to concentrate available resources on a few number of intelligence-gathering operations deemed to be of high importance for one reason or another – in our counter-terrorism context, persons of interest and their activities. But even this more focussed and smaller-scale approach as compared to NSA and GCHG allows the BND to collect about 220 million sets of metadata each and every day, using NSA's *XKeyscore* software for its attempt to achieve digital 'Totalüberwachung' or 'total surveillance' (Spiegel

Online 2013), and passing them on to allied services such as CIA and, of course, NSA (Biermann 2015). Regarding metadata, this means information about who communicated with whom, when, where and for how long. But, as Biermann states, "[whoever] can interpret [metadata] knows not only what people are telling each other, as metadata betray much more: exactly where people are, where they came from, what they are doing at that moment, even what they are planning. They uncover every hiding place and every secret contact" (Biermann 2015). In case the implications of this are not clear, Biermann also quotes former NSA and CIA director General Michael Hayden as saying that "[we] kill people based on metadata" (ibid.).

Even if we ignore the targeted killings carried out abroad (mostly via drone strikes, for example in Afghanistan, Syria and Somalia) in our context of protecting our homelands and our cities, the implications for civil liberties as a by-product, or 'collateral damage', of such vast data-mining operations obviously are far from trivial – I shall come to them below. On the other hand, and again in the context of counter-terrorism and, especially, nipping threats in the bud, the importance of operations such as Prism and Tempora also are far from trivial: scores of potentially devastating acts of terrorism could be detected well before the operators involved were ready to strike: as we remember, we cannot possibly avoid digital footprints when we make use of modern communications technology – and neither can terrorists if they have an attack in mind that involves a) several individuals, and b) any weapons that either need to be sourced first or assembled. One excellent example in this regard is the *Transatlantic Airliner Bombing Plot* of August 2006 that revolved around fairly sophisticated liquid bombs, consisting of highly volatile TATP and disguised as soft drinks in their original and seemingly unopened plastic bottles. No less than seven flights from London Heathrow International Airport to various destinations in the USA and Canada had been selected as targets (five Boeing 777 and two Airbus 380), and no less than 18 suicide bombers would have taken part in this audacious plot. Had the plan come to fruition and succeeded, more than 2000 people would have died. Hence, when we discuss the infringement of civil liberties later on, this should be kept in mind.

There is however another important aspect to data-mining that needs to be discussed: data sharing, or intelligence sharing. As the 9/11 Commission report highlighted, all knowledge discovered by one government agency may come to naught if it is not shared with other agencies and services for which it might be of relevance – something which is hard to decide a priori. This implies that data-mining efforts also require a certain amount of data-sharing efforts to maximise results. Apparently, this is done: NSA, CIA, GCHQ and the BND share their information, at least to a certain extent. Information sharing however can be quite an uphill struggle even within domestic borders: experience shows that government agencies, especially intelligence services, tend to jealously guard their data for a variety of reasons, some sound and some not so sound. The protection of one's sources, including human sources, plays a big role in that regard, and can be seen as a valid excuse. On the other hand, the reluctance to share information with other agencies oftentimes boils down either to institutional inertia or to a simple 'no need to know, no need to share' stance. At times, the situation is aggravated by turf wars, and by attempts to get a bigger slice of an ever-shrinking government budget.

Occasionally, Murphy's Law stands in the way of effective information sharing: all that is needed that crucial information remains unshared is that the name of a person of interest is entered differently in different databases. This is a problem mainly when it comes to non-Western origin names that need to be transcribed first. An example for that is the name of the erstwhile leader of Libya Muammar Gaddafi: while his personal name was also transcribed from Arabic to English as Moammar, Muammer or Mu'ammar, his family name was Romanised either as Gadhafi, Gathafi, Khadafi, Khadafy, Qadhafi, Qaddafi, and Qadhdhafi, with or without diacritics, while he himself seemed to have signed as Moammar El-Gadhafi. And these are only the most frequent transcriptions – the Library of Congress has collected many more of them (Library of Congress, undated). Similar problems exist with the transcription of Chinese names, aggravated by the fact that there are the famous 'old hundred surnames' which are the most frequently used ones – thus resulting in hundreds or even thousands of persons with the same name. On the other hand, the old Javanese habit of adopting just one personal name instead of a combination of personal name and family name, for example Sukarno or Suharto, also results in hundreds of false positives in a name search.

But even if everything goes as planned, it might turn out that in the huge haystack of data that has been painstakingly sifted through, there simply was no needle. An example from Germany is given by Amoore and de Goede, who first describe the sheer scope of the data-mining effort undertaken in the immediate aftermath of 9/11 to discover potential 'sleeper cells' hiding in German cities – an effort that involved data 'harvested' (that's the technical term for the process of data collection) from no less than five million males:

> Data were collected from universities, private companies, municipal authorities and immigration authorities by a set of criteria that included sex and country of origin. The objective of the process was to find connections between otherwise insignificant pieces of data that in combination were thought to be cause for suspicion. [...] The German search did not yield any terrorist suspect or prosecution. (Amoore and de Goede 2008, 7)

The only tangible yield of the data-mining effort was a rather unexpected, but probably entirely foreseeable one: a Moroccan student objected against the harvesting of his data and went to court. He won his case: "in 2006, the German High Court ruled [the data-mining effort] unconstitutional" (Amoore and de Goede 2008, 19). And with that, it is time to take a look at data-mining's impact on civil liberties.

6.4 Evaluation: Data-Mining and Civil Liberties

Whatever the merits of data mining for the purpose of counter terrorism may be, here are obvious civil liberties implications here, as Fienberg emphasizes – especially when it comes to the predictive goal of data mining that involves the sifting of vast amounts of data of perfectly innocent (one hopes) citizens with a 'no holds barred' approach and the believe "that every database and every source of information is fair game – airline and other transportation records, telephone call logs, responses to surveys, credit card purchases, census data, etc. – irrespective of promises of protection, legal and otherwise" (Fienberg 2008, 198). Indeed, data mining in general and programs like the aforementioned *Total Information Awareness* met with stiff resistance from the American Civil Liberties Union (ACLU) that also protested again the USA Patriot Act as such, highlighting that the Patriot Act "[hastily passed 45 days after 9/11 [...] made it easier for the government to spy on ordinary Americans by expanding the authority to monitor phone and email communications, collect bank and credit reporting records, and track the activity of innocent Americans on the Internet [thus turning] regular citizens into suspects" (ACLU, undated; also see Siegler 2006). And in the UK, the so-called *Snoopers' Charter* or Investigatory Powers Bill of 2016, as its official name goes, also resulted in protests from various human rights bodies because it is deemed to be an unlawful bulk data collection 'just in case' that also includes data sets of all those individuals "that aren't suspected in any wrongdoing but have been swept-up in the data collection" (Burgess 2017). This collection of data is carried out not only passively by sifting through available information, but also actively by way of hacking "into computers, networks, mobile devices, severs and more," targeting both UK citizens as well as foreigners abroad who, for one reason or another, "have come to the attention of the security services" (Burgess 2017). Furthermore, to enable investigations of intelligence and police forces, communications service providers are now required to store Internet Connection Records or 'meta data' containing crucial information about "who, what, when, and where" for twelve months (ibid.).

Some fears of concerned citizens regarding a breach of their privacy are obviously valid: being swept-up by an investigation without one's knowledge certainly is a cause for concern and should not simply be swatted away with the usual platitude of 'if you don't have anything to hide, you don't have anything to fear.' After all, the risk of 'false positives', already discussed when examining biometrics also is an issue for data mining – and a far more serious one than that created by an incorrect facial or iris scan that usually can be resolved rather quickly via a manual check done by a human operator: ending up on a terrorist watch list or on an anti-money laundering one without realising it does occasionally happen, but getting off such a list is well-nigh impossible, as some unfortunate innocent individuals found out over the years: since the risk assessment calculation that resulted in someone been included on a watch list is classified (as described above in the case of ATS), there is not much chance to appeal against the inclusion. Hence, as Fienberg points out, it is not the postdictive identification of known terrorists that lead to human rights concerns but the predictive one, since this kind of 'dataveillance' (Clark 1988, 498) involves the mining of immense amounts of data of perfectly legal users without their knowledge or consent – and of course with the possibility of falsely identifying innocents as terrorists (Fienberg 2008, 203).

But the sheer amount of data collected in such very broad bulk data collections renders it very unlikely indeed that those whose data are just swept-up as a kind of 'collateral damage' would have to worry about members of intelligence services or police forces having a laugh for example over their drunken pictures from the last holidays which they put on Facebook or Instagram: even the biggest services such as the NSA do not have the personnel for that, and most certainly not smaller services such as the BND in Germany. Furthermore, there are clear protocols and codes of practice that need to be followed when it comes to such operations, and there are regulatory bodies, judges and ombudspersons involved as well – hence, as already argued, it is not simply a 'free for all', the odd case of misuse and abuse notwithstanding. Some of the more far-sighted states, or at least those prodded on by their various supreme courts, already have included such safeguards even before modern data-mining became feasible. The German Federal Constitutional Court for example did so as early as 1983 as a reaction to wide-spread protests against the census (Volkszählung) of that year which were also infused with a certain unease regarding the BKA's data collection activities starting in the 1970s (as discussed above): in its judgement, the Court created a 'constitutional right of informational self-determination' (Grundrecht auf informationelle Selbstbestimmung) which is meant a) to guarantee the protection of individuals against unlimited collection, storage, usage and transfer or disclosure of personal data, and b) to guarantee the right of every individual to decide for himself or herself how personal data (in an Anglo-American legal context: Personally Identifiable Information, PII) should be used or disclosed under the conditions of modern data processing. More recently, in November/December 2017, the UK introduced new safeguards and amendments to protect at least its own citizens against an unrestrained collection of communication data (Travis 2017).

Such safeguards and amendments notwithstanding, it can still be argued against the backdrop of Big Data and surveillance programmes such as NSA's *Prism* and GCHQ's *Tempora* that domestic laws guaranteeing informational self-determination are more or less obsolete by now since they cannot possibly protect the state's citizens against surveillance programs run by other states' agencies who could not care less about such 'foreign' (from their perspective) constraints (see for example Franzius 2015). Possibly in the light of this, and probably also irked by NSA's Prism programme, the European Union itself took action to protect its citizen's Personally Identifiable Information by updating its meanwhile outdated directive on such matters of 1995. On 27 April 2016, and after a relatively short period of deliberation, Brussels passed a law regulating data protection, titled *EU Data Protection Regulation,* or EUDATAP for short (EUR-Lex 2016). After a period of two years (during which the member states were meant to incorporate its provision into domestic law), EUDATAP became part of EU law in May 2018. How viable EUDATAP's protection will be, especially against large-scale data-mining and surveillance measures of non-European intelligence agencies, remains to be seen at the time of writing. Regarding Prism for example, the Council of Europe's commissioner for human rights Nils Muižnieks complained that "secret, massive and indiscriminate surveillance programmes are not in conformity with European human rights law and cannot be justified by the fight against terrorism or other important threats to national security" (as quoted in Bowcott 2014).

Again, this may indeed be deplorable when seen from a civil liberties perspective, but if we stay with our theme of technology as a silver bullet in the fight against terrorism for the moment, it is clear by now that data-mining, especially real-time data-mining, is a far too important weapon to be dismissed, or to be encumbered by too many legal obstacles. Some of the most ambitious of scholars, politicians and leading police officers even dream of a *Minority Report*-style prediction of crimes yet to be committed. In the People's Republic of China for example, Li Meng, Deputy Minister for Science and Technology, is convinced that in the near future, advanced data-mining based on artificial intelligence (AI) and machine learning (ML) will help Chinese security forces to prevent terrorist attacks or crimes via a much more precise predictive profiling than that previously possible (Strittmatter 2018). Whether real-time data-mining, especially a predictive one, is a price worth to pay, and whether it would be possible to stop such practices in the first place now that they have been introduced, is something we will have to discuss in the third part of this book.

The irony here is that much of this large-scale data-mining by various intelligence services is made possible not by an enforced total observation run by an Orwellian 'Big Brother' state, but by a network of 'Little Brothers' in the shape of big companies such as Google, Yahoo, Facebook and others, who we voluntarily surrender very private information about us via our smart phones, fitness trackers and other smart gadgets that we use without much reflection, or any second thoughts. As journalist Adrian Lobe points out, surveillance does not feel like suppression any longer but like fun (Lobe 2017). Or, in the words of Randolph Lewis, surveillance does not look like a *Panopticon* any longer, but more like a *Funopticon* (Lewis 2017, 52). And the frequent travellers among us should also not forget just how convenient it can be to be (voluntarily) enrolled on one of the already existing *smart border* programmes, a few of which I already mentioned above: beating the queue by effortlessly (most of the times, at least) by walking through an e-gate does have a certain appeal, does it not? Although Matthew Sparke uses it with a rather negative connotation, I admit that I quite like his "cushy cosmopolitanism" (Sparke 2008, 133) as a term to describe the positive experience of being able to do so, and in being, obviously, a 'trusted citizen' – the latter being a Chinese moniker with decidedly sinister connotations from a liberal Western perspective, which I will discuss later on in the third part.

Chapter 7
Detection: Scanning and 'Sniffing' Technologies

Abstract Even the best efforts of counter-terrorism specialists sometimes come to naught when an individual or a group with ill intent slips through the net for one reason or another. Hence, in this chapter I look at yet another 'silver bullet' in the shape of technology that enables us to see beneath the layer of clothing, or beneath the canvass of a backpack, in order to decide whether there is a threat or not – especially in the vicinity of access points of locations where many people congregate, thus offering an easy-to-attack soft target. The technologies in question are metal detectors, full-body scanners, explosives detection systems, and explosives trace detectors. I also discuss the advantages and disadvantages of various systems, in particular highlighting the controversies around the first generation of full-body scanners.

Keywords Metal detectors · Full-body scanners · Automated Threat Recognition (ATR) software · Explosives detection systems · Explosives trace detectors

Even the best efforts of counter-terrorism specialists sometimes come to naught when an individual or a group with ill intent slips through the net. Maybe they could, against all odds, keep the planning and reconnoitring phase of their intended attack out of sight of even the most advanced data mining systems. Maybe they plan to act alone, using a car or van to launch their attack, or even just a knife – which makes it well-nigh impossible to detect their intent to strike before they do so because there are no tell-tale signs to be found: after all, no weapons need to be sourced from the Dark Web, no precursors of explosives bought, such as peroxide. Or maybe they are just some 'clean skins', that is persons who never raised any suspicion, and never appeared on the counter-terrorism radar screen in the first place. This means that our next chance to stop these individuals is by detecting them before they are ready to strike even though they might already be on the way to the intended location of the attack, maybe carrying explosives on his or her body, or being armed with guns or knives and the like. Basically, what we are looking for is yet another 'silver bullet' in the shape of technology that enables us to see

P. Lehr, *Counter-Terrorism Technologies*, Advanced Sciences and Technologies for Security Applications, https://doi.org/10.1007/978-3-319-90924-0_7

beneath the layer of clothing, or beneath the canvass of a backpack, in order to decide whether there is a threat or not – especially in the vicinity of access points of locations where many people congregate, thus offering an easy-to-attack soft target. These are the technologies Mills and Huber (2002, 1–2) refer to when they argue that "we know we have to see the plastic explosives in the truck before they detonate, the anthrax before it's dispersed, the sarin nerve gas before it gets into the air-conditioning duct – and not just see but recognize it. Our imaging systems will have to distinguish between the scaffolding on a Wonder Bra and the wiring on a bomb."

To a certain extent, modern technologies already allow us to look beneath the layer of clothes, and even beneath the skin. Due to the terrorists' ongoing fascination with aviation (Price and Forrest 2016, 1), airports were (and still are) at the forefront of the testing and the deployment of such technologies. As Sorkin (2004, 261–262) puts it, the average international airport nowadays "is the primary training ground, the vanguard of the organisation of the city as a space of heavily surveilled, highly managed flows [...] as well as the premier research site [...] for the burgeoning technologies of surveillance, data processing, perpetual motion, information management, and other top-down styles of security." It thus makes eminent sense to assess the range of detection technologies developed for and deployed at airports, to then see whether they could be used for general counter-terrorism purposes as well, especially with regard to protecting our open societies and our cities, our critical infrastructures and our key assets.

7.1 Metal Scanners: About Portals and Wands

The first devices introduced to enhance aviation security were metal detectors in the shape of portal scanners and portable scanners. Portal scanners, known as *Walk-Through Metal Detectors* (WTMD) or magnetometers for obvious reasons, detect metals by creating a magnetic field, which is interrupted when someone walks through with something metallic hidden on the body. The WTMD then triggers an alarm which usually results in a thorough pat-down by security personnel who may also use a hand-held metal detector, colloquially known as a 'wand.' As frequent travellers know from their own experience, walking through the portal takes a second or two – but at some airports that have chosen to dial up the sensitivity (which can be adjusted) of the WTMD to its highest setting, this initial walk-through is oftentimes followed by a repeat after the shoes had to be removed and put on the belt as well to also be scanned by a smaller portal scanner – just like one's carry-on baggage (the system doing this will be described later on). In itself, the procedure is not too onerous and certainly not invasive, but it can be quite annoying when trying to get to a connecting flight with only a short layover at a busy airport with massive queues. Hopefully, this will be made redundant by the introduction of newer WTMD systems that, unlike earlier models, do not only give an alarm when metal is detected but also point at the location of the object. Price and Forrest for example describe a

WTMD system, the Garrett PD 6500i, that "divides the magnetic field into 33 distinct zones and has uniform coverage from head to toe. Lighting visible on the exit side of the detector notes the zone where a particular metallic object is located" (Price and Forrest 2016, 311).

The comparatively cheap deployment costs (about US$30,000 per unit) and user-friendliness of WTMD systems, both for operators as well as for those having to step through them, resulted in their swift and widespread adoption for access control of official as well as private sector facilities. Regarding the former, WTMDs often are part of the access control of certain government buildings that, due to their function, are seen as permanently at risk of being infiltrated by individuals with not necessarily the right kind of intention, be it terrorist or criminal. Western embassies for example always require visitors to step through portal scanners, on top of other security measures. In a time of heightened risk of terrorist attacks, many shopping malls have adopted WTMD systems as well, also checking bags if and when required – although this practice is not always taken as seriously as it probably should be. Nevertheless, it renders it at least a little bit more difficult to smuggle weapons or improvised explosive devices (IEDs) into the premises and might thus entice attackers to look elsewhere for easier targets (an effect known as 'threat displacement') – not that such detectors alone would impress or slow down perpetrators of marauding attacks like the ones in Mumbai and in Paris. But still, deploying these systems definitely is a step in the right direction.

But even with the most-modern WTMDs, the main problem with metal detectors is exactly that: they detect metal and nothing else. This means that in a time of a changed threat environment, for example in the aftermath of the Transatlantic Airliner Bombing plot of 2006 that revolved around liquid bombs and the Underwear Bombing plot of Christmas Day 2009, these systems turned out to be insufficient: neither of these highly sophisticated terrorist devices would have been detected by a metal scanner – as demonstrated by the fact that the 'underwear bomber' Umar Farouk Abdulmutallab could board his connecting flight to Detroit at Amsterdam Airport Schipol, and before that, his flight from Lagos to Amsterdam. Also, ceramic or nylon knife blades would remain invisible for such systems. To echo Mills and Huber, metal detectors would neither see nor recognize such devices. The threat posed by such devices may well be exaggerated and far-fetched as Klitou (2008, 318) argues, but for authorities responsible for aviation security, it was plausible enough to call for a better system. These calls were further fuelled by the apparent inability of security personnel to always detect those fake explosives, pistols and knifes that undercover testers, or occasionally even journalists, carried on their bodies, hidden under their clothes (Klitou 2008, ibid.). So-called *Full Body Scanners* using Advanced Imaging Technology (or AIT for short) promised to fill this perceived security gap.

7.2 Full-Body Scanners: 'Looking Good Naked' in a Different Context

As Price and Forrest explain, AIT technology comes in two variants: backscatter X-ray technology (or backscatter technology for short) and millimetre wave imaging technology. Regarding the former, backscatter technology "projects low-level X-ray beams over the body to create a reflection of the body displayed on the monitor" (Price and Forrest 2016, 312). Regarding the latter, millimetre wave imaging technology, the level of radiation individuals going through such scanners is lower than that emitted by mobile phones and thus deemed to be safe. As Price and Forrest put it, such a system "bounces harmless electromagnetic waves off the body to create the same generic picture for all passengers" (ibid.) – a rather crucial detail to which I shall return below. At the moment, it is more important to briefly highlight the disadvantages that both technologies share. First of all, compared to the WTMDs, full body scanners are far more expensive with an average cost of US$250,000 per unit or more, depending on the model. Secondly, the whole scanning procedure takes up to 45 seconds, during which the individual needs to stand still, with arms elevated above one's head. Compared to WTMD, 45 seconds is quite a long time – which means usually not everybody is required to walk through these scanners, at least as long as the threat of terrorism is not deemed to be highly likely or imminent (the corresponding two highest threat levels of the US Homeland Security Advisory System are 'high' and 'severe'; the two highest UK threat levels are 'severe' and 'critical') – usually, individuals are randomly selected for this extra level of scanning, apart from the metal detector. Furthermore, since both systems either cannot penetrate the skin (millimetre wave technology) or only about 2.5 cm or 0.1 inches (backscatter technology), items hidden in body cavities or in shoes are essentially undetectable.

Initially, there was however one much more fundamental problem with the new backscatter body scanners I already hinted at: as Klitou (2008, 317) points out, "backscatter body scanners, by design, also intrude upon the privacy of the human body." Klitou refers to the fact that the first-generation systems deployed in the USA from early 2007 onward produced a rather detailed greyscale picture of the person in the scanner – a very detailed nude greyscale picture, to be precise, which, after a simple colour inversion, looked even more revealing and life-like. Indeed, the early systems even "show[ed] nipples [and] the clear outline of the genitals" (Bill Scannell, as quoted in Klitou 2008, 317). Furthermore, the first-generation systems could "also reveal sensitive medical details about a person, such as mastectomies and colostomy appliances" (Klitou 2008, 317). Hence, the process of going through a backscatter body scanner was likened to a 'virtual strip search' by civil rights groups such as the American Civil Liberty Union (ACLU) (Klitou, ibid.). Not surprisingly, it did not take long for the full body scanners to produce negative headlines due to their massive assault on people's privacy and dignity. The *Economist* for example reports an incident in the USA of May 2010 in which "a Transportation Security Administration screener [was] facing an assault rap after he allegedly beat a co-worker who joked about the size of the man's genitalia after he walked through a security scanner" (Gulliver 2010).

Under the impression of Umar Farouk Abdulmutallab's attempt to blow up an airliner during its approach to Detroit at Christmas Day 2009, full body scanners of the millimetre wave variant became compulsory in the UK from 1 February 2010 onward. The first two airports to deploy such scanners were London Heathrow International Airport (LHR) and Manchester Airport (MAN), with six other British airports following suit within six months. Eleven more British airports deployed full body scanners from 2014 on. Outside of the UK, Finland tested full body scanners at Vantaa Airport Helsinki, and the Netherlands at Amsterdam Airport Schipol – Schipol being one of Europe's most important aviation hub. In 2010, France did the same at Charles de Gaulle Airport in Paris, and Italy at the airports of Rome and Milan (EUR-Lex 2010). Just like in the USA, the reaction of the public first was somewhat bemused, then decidedly negative. Very quickly, a spontaneous and largely unorganized grassroots protest movement spread across Western Europe, targeting some of the airports that had already introduced the systems as ordered. Some of the protests had a funny side to it, such as one in Frankfurt (which actually did not participate in the testing of such systems) in which several dozen young people in their underwear entered the departure hall with sarcastic slogans daubed on their bodies. Other protests were of a more serious nature: several child protection groups raised the alarm that paedophiles could abuse the system to take and collect nude pictures of underage persons, while others feared that paparazzi could bribe airport employees to make nude pictures of celebrities going through the scanners. Although most of these fears appeared to be over-blown, and although both the firms producing such scanners and airport officials quickly asserted that it would not be possible to store pictures to access them later, controversies were not long in coming. Just one month after the introduction of these full-body scanners at Manchester, in March 2010, the first two passengers who refused to go through them were denied boarding their flight. Said passengers were Muslimas who believed that the scanners violated Islamic teachings on nudity. In the same month, an airport security officer was given a harassment warning by the police because he allegedly took a picture of a female colleague walking through a scanner. Both cases reignited the debates, with the first one adding a 'religious discrimination' flavour to it as well.

Since then however, the technology has been adapted – in the UK in the wake of a public consultation in which all the concerns mentioned above were flagged up. As a result of this, a 'Security Scanners Code of Practice' was published on 21 November 2011 (latest version October 2016) that stipulated that the privacy of individuals had to be protected via an Automatic Threat Recognition (ATR) software. The document described the software and the screening process as follows:

> ATR software interprets the scan data, instead of creating an image, and identifies areas where items may be concealed on the body. These areas are flagged on a standardised stick-figure on a screen, to indicate to the security officer areas of the individual's body which should receive a targeted hand-search [...]. Immediately after the scanning analysis is completed and the individual moves away from the security scanner, all data relating to the

individual must be destroyed, irretrievable and incapable of being copied or sent. (Department for Transport 2016a, 2)

The document further stipulated that passengers should be informed about the necessity of undergoing a scan by a full body scanner at the security screening area, and about the fact that no images of individuals would be created or stored (ibid.). It also includes an alternative for those passengers who still prefer to opt out in the shape of an "enhanced hand search [...] in a private room" (ibid.). Furthermore, the document states that "[individuals] must not be selected on the basis of personal characteristics (i.e. on a basis that may constitute discrimination such as disability, sex, gender reassignment, age, race, religion or belief, pregnancy and maternity and sexual orientation)" (Department for Transport 2016a, 3). It however does not state what the selection criteria should be. Although this could be criticised from a civil liberties point of view, it does leave airports using these systems some discretion to react to changes in the threat environment. I will address these ambiguities in the third part however, since it does indeed enable a certain 'us' versus' them angle to creep in. Regarding the security procedures of British airports, the document now forms part of the Security Scanners Direction 2016 (Department for Transport 2016b).

All taken together, the good news here is that enhanced security does not necessarily require the sacrifice of one's personal pride any longer, even though these adapted scanners still might affect cultural or religious sensitivities or be seen as violating one's privacy – which is why many airports offer passengers the alternative to be patted down – although personally, I cannot recall having even once been alerted to this alternative before I was made to step through such scanners at various regional British airports between 2011 and 2018. The bad news is that even the latest generation of these full body scanners are unable to detect anything hidden in body cavities, or anything that has a low density. Some aviation experts even claim that full body scanners can easily be defeated. Israeli aviation security expert Rafi Sela for example asserts quite bluntly that he could "overcome the body scanners with enough explosives to bring down a Boeing 747 [which is] why we haven't put them in [Ben Gurion] airport" (Julian 2010).

From anyone else, this would sound like grandstanding – however, since Sela has 30 years' experience in aviation security and actually helped to design the security systems and procedures at Ben Gurion Airport (Julian 2010), his warning should be heeded: obviously, full body scanners are not as capable as manufacturers producing them, and airport operators deploying them (due to official directives) would want us to believe. Hence, for some civil liberties activists, the crucial question has not changed much: Do full body scanners violate our fundamental human rights, without contributing to a substantial increase in aviation security? In my opinion, they enhance security to a certain extent, without negatively affecting my sense of privacy – at least no more than being thoroughly patted down. But this is a question that individual passengers have to answer for themselves. From an airport's perspective, probably of more importance given the rather limited budgets they have is the fact that these scanners are rather expensive. This also militates against the use of such systems outside of airports, except maybe for a small number of high-security facilities: the combination of high costs, slow speed, and still lingering-on civil liberties and human rights issues means that full body scanners currently are of no interest when it comes to protecting railway stations, shopping malls, sport stadiums or other venues with the high throughput of people.

7.3 Explosives Detection Systems: Looking for Homemade Explosives

An even more expensive variant of these X-ray-based explosives scanner systems is the *Explosives Detection System* or EDS. Just like the other scanners, it uses computerized tomography to detect explosives hidden in bags or suitcases. At airports, EDS systems were first introduced in order to screen checked baggage (that part of the baggage that goes into the hold of an airliner), and later also for carry-on baggage inspection. The date of the initial introduction varies across countries, and closely corresponds to acts of aviation terrorism: Canada for example introduced the first EDS system for checked baggage after the bombing of Air India Flight 182 from Vancouver to New Delhi and Mumbai in flight over the Atlantic on 23 June 1985 by the Sikh extremist group Babbar Khalsa (all 22 crew and 307 passengers died), the United Kingdom after the bombing of Pan Am Flight 103 from Frankfurt to Detroit via London over the Scottish town of Lockerbie on 21 December 1988 (all 16 crew and 243 passengers died plus 11 people on the ground), and the United States after 9/11. The European Union, via its European Civil Aviation Conference (ECAC) also made a one hundred per cent hold luggage inspection mandatory under the impression of 9/11. For both the USA and the EU, the end of 2002 was the deadline for the introduction of EDS systems for checked baggage screening (Harding 2004, 870). Russia introduced EDS systems comparatively late in 2004 after two airliners had been blown up in mid-flight on 24 August of that year – Volga-AviaExpress Flight 1353 from Moscow to Volgograd (all 9 crew and 34 passengers died) and Siberia Airlines Flight 1047 from Moscow to Sochi (all 8 crew and 38 passengers died). However, these bombings were in all probability not conducted via bombs smuggled into the airliners' holds but by two Chechen female suicide bombers, so-called 'Black Widows', of the Riyad us-Saliheyn Martyrs' Brigade (Kurz and Bartels 2007, 540, 546). In any case, the costs of acquiring these systems are quite high (they are actually more expensive than full body scanners) and do not necessarily stop with the purchase of such a system as well. This can be gleaned from Price and Forrest's comment on the situation in the USA after 9/11:

> After 9-11, airports purchased EDS machines used to inspect checked baggage at the cost of approximately $1 million per unit. Some major airports had to purchase numerous EDS machines to meet baggage screening demands. Terminal facilities had to be modified and reconstructed to make room for the new systems. (Price and Forrest 2016, 318)

Modifying or even reconstructing terminal buildings became necessary because these scanners had to be integrated into the belt system that transports checked luggage to the right location so that it (hopefully) embarks on the same plane as its owner. To ease the financial burden, in the case of the USA the government promised to reimburse airports struggling to meet the new requirements of the post-9/11 environment. However, the government was rather slow to honour that promise: as Price and Forrest point out, "by 2006, the government had reimbursed about eight airports (out of 429) for EDS upgrades, and additional reimbursement have been slow" (Price and Forrest 2016, 319). Nevertheless, like airports in other countries that had adopted EDS systems before the USA, these costs had to be shouldered – not only by airports, by the way: ocean terminals serving cruise liners that also handle checked baggage also introduced such systems in the aftermath of 9/11 for fear that terrorists would take to the sea to commit acts of 'maritime terrorism.' After all, a major, top-of-the-line cruise liner accommodating about 5000 passengers and crew could also be a tempting target of choice by terrorists (on that, see Lehr 2016).

X-ray-based EDS systems (on the various X-ray techniques, see for example Wells and Bradley 2012, 1737–1742) can detect improvised explosive devices via their geometry and their material density. Regarding the former, the shape of an improvised explosive device (IED) plus its detonator, associated wires and batteries is detectable via an image shape analysis. This simply is a case of 'seeing' and 'recognizing' a threat, to echo Mills and Huber (2002, 1–2). Regarding the latter, the density of the explosive also stands out: it lies between the low density of organic material and the high density of metals (Singh and Singh 2003, 34). However, there are problems with X-ray-based scanning for explosives, as Singh and Singh point out. Firstly, current IEDs are difficult to discover just by an image shape analysis. Plastic explosives such as Semtex or C-4 for example are highly mouldable (hence colloquially known as 'putty explosives') and can take any shape – which means devices based on them are not easily identifiable as a bomb. Secondly, modern detonators for IEDs (or 'Homemade Explosives', HME) are built with only little metal, and wires and batteries also are not that difficult to conceal. As Wells and Bradley alert us, "[illegal] bomb manufacturers seem to be well aware of how X-ray screening equipment works and in as far as it is possible they would probably be seeking to tailor their technology so that the device would become undetectable, particularly in regard to the familiar X-ray transmission imaging systems" (Wells and Bradley 2012, 1737). Thirdly, EDS systems also have a high rate of false positives (up to 30 per cent for some systems) that then require a human operator to be resolved. This human intervention can take the shape of a physical inspection of the baggage and/or an additional swab test by a portable explosives trace detector (ETD) system, which will be discussed below.

In the European Union, checked luggage that fails the initial (Level 1) screening by an X-ray-based system has to undergo a Level 2 screening that consists of a human operator interpreting the image derived by the Level 1 screening at a workstation, with the baggage still continuing its transit on the belt. If the baggage is not cleared by the human operator, it is sent to Level 3 screening that uses different

X-ray technologies (mainly CT scanners) plus human operators for the interpretation of the results. If the suspicious baggage is still not cleared, it has to be reconciled with its owner, and then manually searched. The highest screening level is Level 5, in which the suspicious baggage is considered a bomb threat. As Price and Forrest state, "[a] bag reaches Level 5 classification when the passenger cannot be found or refuses to open the bag. At this point, either the area is evacuated, or the bag is moved via a mobile blast containment system where it is then inspected further or destroyed" (Price and Forrest 2016, 319). According to Price and Forrest, this "occurs on one or two bags in 50 million" (ibid.). In any case, at the time of writing, it is rather unlikely due to the huge costs involved that any of these EDS systems would be deployed outside of airport or ocean terminals/container terminals or a few high-security locations. Hence, like full body scanners, these systems are of limited use when it comes to protect key facilities of our cities. Until the costs are down (don't hold your breath), and until the systems are much faster and much more reliable, they are quite unlikely to make a major impact in that regard.

7.4 Sniffers: Tell-Tales of (Impending) Disaster

Since we are already talking about bombs, I would like to add that despite of what Al Qaeda and ISIS want to make us belief, trying to mix together various volatile ingredients as homemade explosive devices in the kitchen is a rather dangerous venture, and 'results may vary', as one of the most famous disclaimer of pharmaceutical products has it. Sure, the devices might work all too well, as they certainly did in the case of the London 7/7 bombings, or the Boston Marathon bombing of 15 April 2013. They might however miserably fail as well because the mixture was not quite right. This is the lesson of the attempted rerun of London 7/7 two weeks later (in an event that would have then be called London 21/7, had it been successful), in the already mentioned failed *Tiger Tiger* Club bombing of 29 June 2007, or the failed London Underground bombing of 15 September 2017 (on that, see for example Booth et al. 2017). Another possibility is that the device detonates prematurely, for example while being transported to the planned location of the attack – there are some examples of that, too, such as that of a British-born ISIS suicide bomber who in February 2017 accidentally detonated his suicide belt moments before he reached his targets in the shape of a group of Syrian soldiers in the beleaguered ISIS capital Raqqa. A third possibility is that the ingredients blow up when they are mixed – after all, we are talking about very volatile chemical agents here. One example would be the chemical kitchen fire in Manila on 6 January 1995 that led to the discovery of the so-called *Bojinka Plot,* a plot that involved assassinating Pope John Paul II while on visit in the Philippines later that year, and the bombing of eleven or twelve Transpacific airliners in flight from various Asian airports to the USA's Westcoast with liquid bombs. Another one would be yet another kitchen fire in an apartment block off Sukhumvit Road in Bangkok on 14 February 2012 that resulted in the arrest of two Hezbollah members who had planned an attack on Israeli tourists in Khao San Road.

In all these cases, the planned attacks were thwarted not so much by successful counter-terrorism operations, but as a result of Murphy's Law, according to which everything that can go wrong will go wrong – after all, as argued in the introduction, this law applies to the terrorists as well. On many other occasions, such attacks could be prevented due to certain tell-tales that alerted intelligence forces and/or police to the presence of bomb makers in their cities. First of all, in a much more terrorist-aware environment, the attempt to buy certain precursor chemicals needed for homemade explosives in bulk will not go unnoticed – with the most likely consequence being an unannounced early-morning visit by the police. Secondly, and this is of relevance in the present context, certain chemicals required for the production of an IED are rather aggressive by nature, for example bleaching the bomb maker's hair due to their toxic fumes and killing off the plants in the vicinity of the improvised bomb manufacture. In addition, there might also be a rather distinct chemical smell as yet another tell-tale that something untoward is happening. Furthermore, the chemicals needed for IEDs also leave their traces or 'fingerprints' – mostly on the body of the bomb maker himself of herself, but also of everybody else handling the bomb. Logically, this realisation led to the question whether sniffer devices could be developed that could detect these chemical 'fingerprints'. These systems are known as *explosives trace detectors* (ETD), or as 'sniffers' or 'itemisers.'

Again, airports were at the frontline when it came to deploy the first crude sniffing devices. I am sure that some of the frequent flyers among the readers, at certain occasions and at certain airports, had to hand over their laptops to a security official during the security check (or at the departure gate for certain flights) so that he or she could briefly rub its surface with a reagent-impregnated swab, which was then taken to the analyser unit to test for traces of explosives. That security personnel at airports usually are most interested in laptops, tablets and smart phones (especially so since 2017 for flights destined for UK or US locations due to a potential threat from ISIS that explosives-filled laptops or tablets would be used to blow up airliners in mid-flight) does not mean that only such smooth metallic, plastic or glass surfaces can be tested – basically any surface can be swabbed this way, including cloth. Depending on the ETD system in question, trace, bulk, or vapour detection techniques are then used in order to detect and identify traces of potential explosives. At the moment, most commercially available ETD systems are based on ion mobility spectrometry (IMS) or on gas chromatography. As Caygill, Davis and Higson explain regarding IMS,

> IMS characterises a sample through the mobility of ions within the gas-phase of the instrument whilst an electric field is applied [...]. The sample vapours are ionised at atmospheric pressure before introduction into the drift tube. The drift times are related to the mass of the ions and by determining the mass/charge ration, it is possible to identify components within the sample through comparison with known standards (Caygill, Davis and Higson 2012, 4).

Regarding gas chromatography, this technique either uses chemiluminescence detectors or electron capture detectors. The first step for both is identical: samples of materials are taken with a hand-held collection unit "that heats up the sampled surface with infrared lamps and sucks air from the near surface. The sample is then injected into chromatographic columns, which consist of thin tubes lined with a material to absorb molecules of interest. Different molecules are dissolved to different extends and the material with the least affinity for the column substrate goes fastest. Since different materials are released from the column at different times, mixtures of materials can be identified." (Singh and Singh 2003, 44). In the case of chemiluminescence detectors, the molecules escaping from the column are heated, and their chemiluminescent reaction analysed. In the case of electron capture detectors, the escaping molecules are mixed with electrons of a gas mixture ionised by a small radioactive source to test their reaction: according to Singh and Singh, "since the explosives molecules are electro-negative, they grab some of the electrons. Fewer electrons now migrate to the anode and this results in a decreased current. Microprocessors analyse this change for finding explosives" (Singh and Singh 2003, 45). The list of detectable explosives-related substances depends on the method used but typically includes PETN, Potassium, RDX, Saltpetre, Semtex and Nitro-glycerine.

The advantage of such systems is that they are comparatively cheap (typical unit costs are on the low tens of thousands of US dollars, Euros or British pounds), and user-friendly in the sense that their operators can learn how to use them and how to interpret the results in a couple of hours. Another advantage is that basically any surface can be swabbed, as already pointed out. In the context of counter-terrorism, this means that clothing, belts, shoes, bags and backpacks can be tested for traces of explosives, or mobile phones, tablets and laptops – anything that an individual might wish to take through a checkpoint. Furthermore, unlike the stationary X-ray or millimetre-wave-based EDS systems, the portability of ETD systems makes it easy to deploy them wherever they are needed to respond to a certain threat – not only within airports at departure gates of flights deemed to be under threat, but at any checkpoint, even temporary ones set up to meet a heightened demand for security. Major sports events or large-scale open-air concerts readily come to mind in this regard.

Like with all other technical solutions discussed so far, there are some disadvantages as well. First of all, portable ETD systems cannot handle large numbers of customers due to the limited number of security personnel and the time-consuming procedure which is not much faster than a manual pat-down (this means that, depending on the system, about 20 to 30 samples can be taken in one hour). Hence, such portable ETD systems are useful mainly for the screening of more or less randomly selected individuals and their hand baggage or other items (Caygill, Davis and Higson 2012, 3). Also, many ETD systems work best in a controlled environment that is not contaminated by a multitude of vapour traces from outside sources. In this regard, it is fair to add that not all currently available portable ETD systems are actually rugged enough to be used in cold or otherwise inclement weather, and some of them are a bit on the heavy and cumbersome side of things – although these are obstacles that could easily be overcome by the manufacturers if there were a bigger demand for such devices. More importantly, detecting tell-tale traces of an explosive hidden somewhere might be hit and miss: due to the low vapour pressure of many explosives, a thorough wrapping or packaging can relatively easily defeat portable ETD scanners – most of which are unable to detect traces of peroxide (which is used for many explosives) due to the fact that "current chemical identification techniques are based on the nitrogen and carbon content of a substance for identification and this practice is not suitable for peroxide explosives", as Caygill, Davis and Higson (2012, 3) point out. Furthermore, as Singh and Singh elaborate, different detection models come with different limitations: gas chromatography based on electron capture detectors do not detect plastics explosives and other very low volatility materials (Singh and Singh 2003, 45). On the other hand, IMS has problems with detecting "low-vapour-pressure materials such as TNT or RDX" (ibid.). This means that in a high-risk environment, more than one of these systems might have to be used – which prolongs the screening process considerably.

Swab tests also have a certain tendency to deliver false positives that is difficult to eradicate. For example, traces of potassium frequently trigger the alarm since potassium nitrate (saltpetre) is an ingredient of gunpowder (black powder) – but also of certain food groups (for example bananas or tomato products) or of fertilizer. Traces of glycerine also tend to trigger an alarm – understandably so since this is an ingredient in many explosives, for example nitro-glycerine. Unfortunately, glycerine is quite a ubiquitous chemical, and also part and parcel of many cosmetics including liquid soap. Anecdotal evidence has it that at least one unfortunate US traveller was frequently subjected to intensive additional screening and testing because he had washed his bag with liquid soap after an accidental spill – thus unfailingly triggering the alarm every time his bag was checked (Levi 2014). That the unfortunate passenger was not whisked away in handcuffs but just thoroughly screened, padded down and questioned before he could proceed to the departure gate hints at the existence of procedures set up to deal with alarms. The precise procedures pertaining to the so-called 'alarm resolution', that is deciding what exactly triggered the alarm of the ETD device, still are considered *Sensitive Security Information* (SSI) in the parlance of the US Transport Security Administration (TSI) and treated similarly by European Union and British agencies. Hence, it is not allowed for unauthorized personnel (including curious passengers) to watch the monitors of the system, or to take pictures of them. Finally, there might be yet another potential disadvantage of these systems – one affecting the operators themselves. At Munich International Airport for example (which introduced these devices in 2015), a number of security personnel tasked with using sniffers fell ill with symptoms of headache, nausea, numbness, or a general feeling of being drunk. While some of those affected by these symptoms wonder whether these systems might emit hazardous gasses like formaldehyde. However, tests conducted as a result of these illnesses did not reveal anything untoward (Britzelmeier 2018). At the time of writing, it is still unclear if there is a connection between staff members feeling unwell and the sniffer systems or if there is not – at other airports at least, sniffers seem to work as advertised without causing health problems for their operators.

Walk-through portal versions of these EDS systems, known as Explosives Trace Portals (ETP) also exist, either using Ion Mobility Spectrometry (IMS) or Mass Spectrometry (MS) technology. Colloquially, these systems are known as 'puffer machines' since they emit several gentle puffs of air blown at the individual standing in the machine in order to dislodge particles from clothing, hair or exposed skin which are then analysed within seconds. An example for an early IMS-based system would be Smiths Detection's Ionscan Sentinel II which can detect RDX, PETN, TNT, Semtex and some other explosives, plus some narcotics including Cocaine, Heroin, Methamphetamine, or Ecstasy (Linker 2014, 380–381); while the Syagen Technology Guardian would be an example for an MS-based solution. Apart from the different detection technologies used, the basic function is the same: an individual enters the enclosed portal, the entry door shuts while the exit door remains closed, the airflow starts, and the samples are collected and analysed. This process lasts about 20 second. The results of the screening are displayed on the operator's monitor. If nothing suspicious is found, the exit door opens, and the individual can walk on. If a suspicious substance is detected, the exit door remains closed, and the entry door reopens instead. The individual will then undergo further screening by security personnel (Linker 2014, 382–383).

The advantage of these systems over hand-held ETD sniffers are obvious: about 60 to 80 people per hour could be screened by just one portal. A disadvantage is the cost of ETP systems: it is about half a million US dollars – which is not necessarily money well spent: unfortunately, as in many other advanced technologies, these ETP systems promised more than they could keep. For the case of the USA, James Ridgeway states that "in a Denver test by CBS in 2007, a network employee was sprayed with explosives and then walked through the airport's three puffers without any trouble. The machines also set off false alarms, and they frequently broke down, leading to sky-high maintenance costs" (as quoted in Gulliver 2010). The same experience has been made by European airports: after having been trialled in several airports including Amsterdam Airport Schipol as one of the major European aviation hubs, these systems have largely been abandoned by now due to their high rate of false positives. Hence, the verdict of Tim Persons from the US Government Accountability Office (GAO) obviously is quite right: "[they] didn't work very well and there was a large cost" (as quoted in Sheridan 2010). In the meantime, they have been replaced by the already discussed full body scanners anyway because these systems are said to be even more versatile when it comes to the detection of suspicious traces or items, and somewhat more reliable – although also much more controversial, as already discussed.

When it comes to general security and the protection of our cities, all these different methods of detecting traces of explosives with their unique sets of capabilities and limitations share one other huge disadvantage: in order to detect traces of explosives, either a swab or an air sample needs to be taken and analysed – which means such ETD or ETS systems are only useful at checkpoints, be they permanent (for example airports) or temporary (for example, major events taking place at sports stadiums and concert halls) and, due to the limits regarding experienced operators, only on a more or less random check basis. Not surprisingly, considerable research and development efforts are currently underway concerning the feasibility of standoff detectors that could 'sniff' traces of hazardous agents including explosives from a certain distance. First of all, this would enhance the security of the security personnel itself. Secondly, and more importantly for general security, a standoff detection could enable timely counter-actions, for example by automatically closing a security gate through which the individual emitting traces of hazardous material intends to pass through. If this security gate were blast-proof, the result in the case of an attempted suicide bombing would not only be denial of access to the targeted location, but also an immediate consequence mitigation since the effects of a detonation of the device (if the suicide bomber were to trigger it) would be contained within the security gate itself. And thirdly, if it were used in conjunction with a smart CCTV system monitoring public space in the vicinity of access-controlled spaces, it would allow a detection of suspicious persons from among a crowd passing within reach of such a system, again making it possible to inaugurate timely counter-measures, for example by sealing off critical assets nearby. Given the propensity of current EDT systems for false alerts, such systems would need to be far more sophisticated and reliable. Otherwise, the numerous false alerts they would trigger would result in a standstill of parts of our cities – or, more likely, in the understandable decision of harried security personnel to simply switch them off. These formidable challenges notwithstanding, some standoff technologies are currently under development.

One of those technologies is the Quartz Enhanced Laser Photoacoustic Spectroscopic (QE-LPAS) Technology. The QE-LAPS technology that Sharma et al. (2013, 44) see "as one of the most powerful standoff technique for homeland security" is said to have a range of up to 25 metres and able to detect "explosives, chemical and biological materials in gas, vapour, liquid, and solid phase and absorbed materials on the surface or any kind of hazardous material" (Sharma et al. 2013, 44). Since this technology uses a focussed laser beam whose backscattered radiations then are collected by an ellipsoidal mirror to be processed by an oscilloscope via a pre-amplifier and amplifier array, it is probably only of limited use outside of a security channel/security gate: obviously, the laser beam can be directed only at one selected target at a time – hence, detecting traces of explosives carried by an individual hiding in a crowd would be impossible. Furthermore, the reach of the system seems to depend on the material being scanned for traces of hazardous material as well as on the nature of that material: Sharma et al. claim that it would be six metres for traces of PETN and DNT for a wooden surface (low reflectivity) and up to 15 metres for aluminium surfaces (high reflectivity), while the range for detecting traces of acetone and nitrobenzene would be 15 metres for wood and 25 metres for aluminium (Sharma et al. 2013, 46). Be that as it may, it is obvious that it will take quite a while until this system is mature enough to be deployed outside of carefully controlled laboratory conditions.

Chapter 8
Surveillance and Observation: The All-Seeing Eyes of Big Brother

Abstract As the marauding attacks of Mumbai and Paris or the various recent vehicle attacks in Nice, Berlin, London and Barcelona demonstrate, many terrorists select 'soft targets' in the shape of unprotected locations where a crowd usually can be expected. But even if they are bound for hard targets, terrorists are likely to move through public space. In order to heighten the security of such public spaces as well, and in order to prevent an impending attack from taking place, surveillance and observation devices could serve as an early-warning system enabling us to initiate counter-measures. Currently, CCTV systems as well as 'Smart' CCTV systems offer us this opportunity – which is why in this chapter, I discuss the advantages and disadvantages of such systems, to then take a critical look at Augmented Reality (Mixed Reality) as an up-and-coming tool in the fight against terrorism (and criminality in general). Some Chinese police forces for example already make use of 'smart glasses' in order to identify suspects just by looking at them. I conclude this chapter with some critical remarks on this 'Minority Report'-style policing of the (near) future.

Keywords CCTV · Smart CCTV · Surveillance drones · Automated Number Plate Recognition (ANPR) · Smart CCTV and biometrics · Augmented Reality (AR) · Virtual Reality · Smart glasses

As the marauding attacks of Mumbai and Paris demonstrate, or the various recent vehicle attacks in Nice, Berlin, London and Barcelona, not all terrorists are keen to attack so-called 'hard targets' protected by security measures that need to be defeated one way or the other. Rather, many terrorists, especially the self-radicalized ones who do not have any experience that would allow them otherwise, select 'soft targets' in the shape of unprotected locations where a crowd usually can be expected. But even if they are bound for hard targets, terrorists are likely to move through unprotected public space – 'unprotected' in the sense of not featuring access gates controlled by security, not necessarily in the sense of a complete absence of police patrols. In order to heighten the security of such public spaces as well, and in order to prevent an impending attack from taking place, it would be good if we could deploy at least some passive security measures that could serve as an early-warning

P. Lehr, *Counter-Terrorism Technologies*, Advanced Sciences and Technologies for Security Applications, https://doi.org/10.1007/978-3-319-90924-0_8

system that would give us more time to initiate counter-measures. Surveillance and monitoring technologies offer us this chance by allowing us to monitor people's movements at certain locations in order to detect anything that seems to be out of pattern, thus hinting at a potential ill intent.

8.1 CCTV: Visions of Orwell's 1984

Regarding CCTV, we can actually take a look at the (literary) past in order to get a glimpse of the future: George Orwell's famous novel *1984*, written in 1949, already envisaged a future in which every citizen of the fictitious totalitarian state was under visual control basically around the clock. Fortunately, the 'real' 1984 came and went without much fanfare and without finding us living in such an authoritarian system. Regarding the surveillance part of Orwell's novel however, 1984 has indeed arrived. A casual observer might miss it, but a walk through the streets, parks and other public spaces of an average British town or city, followed by a stroll through a shopping mall or a department store, brings us within reach of dozens of camera lenses, gazing down relentlessly from poles, walls, or roofs. The stores we visit have their own CCTV systems, and so have our banks (of course), or most of the public buildings we occasionally visit. The privately-operated systems of banks and stores usually are inscrutable: it is well-nigh impossible to get an idea on where they are aiming at on a given moment or whether they are fake or real (there is actually a good number of 'fake' or 'dummy' CCTVs deployed since they are much cheaper) since their lenses are hidden in a semi-sphere of black or mirrored glass, so-called 'dome cameras'. Many others are fixed in their gaze, always monitoring the same spot in a wide-area surveillance mode. Many of the public ones however are remotely controlled by an operator who can, and occasionally will, directly zoom in on us – which still is somewhat unsettling. With the advent of cheap drones, a 'must have' gadget not only for geeks, it is now also quite likely that airborne CCTV systems soon will do so as 'eyes in the sky', quietly keeping tabs on our movement without us necessarily being aware of this surveillance – which in the particular case of police surveillance would be far stealthier than a rather noisy and easy to spot police helicopter that is normally deployed for such aerial observation missions. Whether we like it or not, this is exactly where modern observation starts: with publicly and privately-operated CCTV systems monitoring public spaces.

Initially, video surveillance or CCTV systems were deployed, both by public bodies and private enterprises by the way, to fight and prevent crime, not terrorism. The first systems were installed as early as 1968 in the city of Olean, New York, to monitor its main business street for the explicit purpose of fighting crime. Five years later, in 1973, CCTVs were also installed at the famous Times Square in New York City, for the same purpose. In Germany, reeling under the impact of the devastating terrorist attack on the Munich Olympic Games of 1972, the first CCTV system was fielded in Munich just one year later in 1973 – albeit also mainly with the purpose of fighting crime, not necessarily terrorism. Initially, the United Kingdom was lagging, setting up video surveillance systems more than a decade later in Bournemouth in 1985 and then in King's Lynn in 1987 (on the early UK system, see Norris and Armstrong 1999, 40–59). Although from a criminalistics perspective, CCTV systems were deemed to be successful tools in at least reducing crime, the adoption of such systems largely depended on cultural attitudes, not on technical merit. In Germany for example, and due to the Germans' suspicions towards state surveillance after their experiences with Nazi totalitarianism, video surveillance was met with stiff resistance from within the affected population – much more so than the largely invisible computer dragnets discussed above. Even nowadays, nearly five decades after the first CCTV system entered service, and faced by random acts of terrorism conducted mainly by self-radicalized and self-declared ISIS members, resistance to such 'Big Brother'-style surveillance is still strong within the German public.

On the other hand, in the United Kingdom, video surveillance proved to be far less controversial – probably aided by a few timely and spectacular successes, not necessarily on the front of crime prevention, but of crime investigation and prosecution. For example, the two juvenile kidnappers and murderers of two-year old James Bulger were recorded by a CCTV system based in the shopping mall in which they spotted and took away the unsuspecting toddler on 12 February 1993. With the help of video analytics and the pictures taken by this early video surveillance system (which only had a low frame rate just like the bulk of the systems deployed in the 1980s/1990s), the two kidnappers could be identified, while the still pictures were used as evidence against them later in court. Similarly, video surveillance systems helped in identifying and in bringing to justice David Copeland, also known as the London Nail Bomber, who carried out three bombings between 17 and 30 April 1999. In the context of terrorism, video surveillance systems again proved to be very valuable for identifying the four London 7/7 bombers: they were picked up by CCTV at Luton when they entered the Underground station to carry out their deadly attacks. Two weeks later, CCTV systems also helped to identify the four unsuccessful London 21/7 bombers since they were also picked up on their way to their own terror mission, meant to mirror the 7/7 bombings. As regards the already mentioned London Riots of 6–11 August 2011 (which also spread to Birmingham, Nottingham and Greater Manchester), about 3,100 suspected participants could be identified and arrested – the bulk of them with the help of CCTV footage.

Of importance regarding counter-terrorism and, especially, the aspiration of nipping the threat of terrorism in the bud, is the fact that none of these systems could prevent the attacks – neither the kidnap, nor the bombing spree of Copeland, or the much later 7/7 and 21/7 events. Hence, it is fair to say that despite these high-profile cases, there is not much evidence that video surveillance systems as such would have a measurable impact on preventing terrorism or crime – all these systems seem to be able to do is helping police forces in the aftermath of an event to piece together the puzzle of what exactly happened and who was involved to prosecute them. Unsurprisingly, this triggered a debate around the wisdom of deploying ever more such systems, and the overall effectiveness of them versus civil liberties. The fears of the United Kingdom turning into a surveillance state à la Big Brother hark back to the question I asked in the introduction: how far can we as liberal democracies go to improve our security? After all, it is an open secret that CCTV systems deployed by private and public bodies also work well as means for social control and exclusion – the latter in conjunction with gated, or rather 'segregated', communities in the 'discourse of fear' that I also mentioned in the introduction. As early as 1999 Norris and Armstrong highlighted this issue, arguing in the case of privately operated CCTV systems guarding gated communities that "[in] these 'fortress cities', CCTV is used to police the boundaries of segregated space and ensure swift response to unauthorized entry" (Norris and Armstrong 1999, 9). The fear of an Orwellian 1984 still is a bit overblown however, despite the recent revelations around the just discussed large-scale data-mining operations of intelligence services such as the National Security Agency (NSA) in the USA or the General Communications Headquarters (GCHQ) in the UK. As Norris and Armstrong (1999, 7) remind us, "CCTV has been implemented not as one pervasive system but as a series of discrete, localised schemes run by a myriad of different organisations rather than a single state monolith. There is no single 'Big Brother' who is watching over us but lots of little brothers with their own agendas." This still is the case today, about two decades later. Furthermore, at the moment, it is still not that difficult to circumvent or fool the rather clumsy and technically mostly obsolete CCTV systems that form the bulk of those that we can spot in our cities. Donning a hat or wearing a hoodie usually is enough to defeat the operators watching (or not) the screens. Hence, the adage of 'seeing is believing' does not necessarily ring true in this context.

8.2 Smart CCTV: Detecting More than What Meets the Eyes

Modern CCTV systems, or 'smart CCTV' systems, aim at overcoming these limitations by adding other technologies to the camera lenses as well. Hence, smart CCTV systems are smart in the sense that they go beyond the mere visual inspection of potential suspects by a human operator. Not all of those systems (especially not the earlier ones) are aiming at individuals in order to identify them – rather, they train their camera eyes on vehicles that pass their range of vision. Many of these systems are fixed-view systems installed along certain main roads, for example on fly-overs and signal bridges crossing main motorways, or at major crossings and intersections in cities. Of special interest for these systems are not the vehicles themselves but their license plates, which are automatically scanned as the cars, vans, trucks, busses pass by. The license or number plates are checked against a data base to see whether the vehicle in question has previously come to the attention of the police. If so, an alarm is triggered to alert a human operator. In the United Kingdom, this system is known as the *Automated Number Plate Registration* (ANPR) system. In a counter-terrorism context, it briefly made headlines in the aftermath of the London *Tiger Tiger Club* car bombing plot of 29 June 2007. Due to a faulty set-up, the two car bombs did not detonate – which made the perpetrators flee to Scotland, where they carried out a last-ditch attempt to perpetrate a terrorist attack by ramming a Jeep Cherokee into the arrivals hall of Glasgow Airport on the next day and setting the vehicle ablaze. One of the two perpetrators died in the event, while the other could be arrested. Five other people were slightly injured. What is more interesting in our context is that during the police investigation, an ANPR system on the M6 motorway detected a car belonging to an individual suspected to be involved in the planning of the airport attack. This led to the interception of said vehicle by several police patrol cars, and the arrest of the two persons found inside (BBC 2007).

At the moment, ANPR systems can only read the license plates – which for counter-terrorism purposes is not yet good enough. Hence, discussions revolving around the feasibility of an updated ANPR system centre on the inclusion of facial recognition technology. Such a 'smart' ANPR system would feature two cameras: while on camera would read the number plate and compare it to a data base of license plates that are, for one reason or another, of interest to the police, another camera would zoom in on the driver to compare his or her face with yet another data base containing facial data of so-called 'persons of interest.' Coaffee however cautions that "this type of technology is unlikely to be available at present [and that using] facial recognition technology to 'snap' the driver rather than the number plate of a vehicle would also necessitate different legislation […] and a rigorous code of conduct set for the operators and monitors of the system" (Coaffee 2004, 293). Be that as it may, in July 2016, the governor of New York, Andrew Cuomo, introduced an ambitious plan for a fundamental high-tech makeover of New York's bridges and tunnels – a plan that also included the installation of CCTV systems complete ANPR capability and facial recognition software both to combat crime and terrorism (Moon 2016). Apart from increasing the security of New York's citizens and visitors, this system would also be quite convenient for the average user in a welcome side-effect: for those enrolled in the programme (I assume that an enrolment is required), there would be no need to stop at toll booths to pay one's fee since the system could automatically identify a driver, either by reading his or her face (still challenging at the moment, but technology will get there) or by reading the car's number plate (much more straightforward at the moment).

The planned 'smart' ANPR system leads us to the current cutting edge of smart CCTV systems: the pairing of CCTV with biometrics recognition software, in the hope to detect suspect individuals, for example those on a terrorist watch list, from among a crowd as they pass through the street or reach an entry point to a secured location. Best candidates in this regard are biometrical features that a) are difficult to hide, and b) do not depend on the cooperation of their 'owner' in order to be used by software-based recognition systems paired with high-resolution CCTV – hence, biometric features perfect for unobtrusive monitoring, surveillance and identification purposes, even without the knowledge and the consent of those being monitored, surveilled, and identified. From the list of biometrical features discussed above, only two stand out: one's face, and one's gait. Regarding the former, Introna and Wood state the obvious when they remind us that the "targets do not need to surrender their face image, as they would their fingerprint, or their iris scan. A face can be captured and (de)coded without the consent or participation from those being targeted" (Introna and Wood 2004, 178). The same is true for the latter: an individual's gait also can be captured and decoded without the individual's consent, and also without his or her knowledge. Hence, both technology can well be called 'silent technologies' (Introna and Wood 2004, 183).

Particularly interesting in this regard is Introna and Wood's mention of coding and decoding: as explained in the chapter on biometrics, any biometric feature can be turned into an algorithm, and thus 'coded' as a template. 'Decoding' however implies comparing this algorithm with existing templates for identification purposes – which requires the existence of a database. So far, gait recognition is not yet that widespread, and thus of limited value for the near future. This is however not true for our faces, as Introna and Wood point out:

> Unlike fingerprints or DNA samples, which are only collected when there is a reasonable level of suspicion of a crime, face images are routinely collected in society by a variety of institutions, such as when we apply for a driving licence, or a passport, or a library card, etc. It is the most common biometric in use by humans to identify other humans. Indeed, in any western society, if one would somehow cover, or be seen to attempt to disguise one's face, then there is almost an immediate assumption of guilt. One could almost say that there is an implicit common agreement to reveal our faces to others as a condition for ongoing social order. Thus, we tend to reveal our face to others and they to us. (Introna and Wood 2004, 178)

There are limits, of course, as Introna and Wood also point out: we do not necessarily like it if and when strangers take photos of us without asking for permission first, to put them on social media for everyone to see. In the USA for example, a former Playboy playmate ended up in court in November 2016 for taking a picture of a 70-year-old woman at a gym without her knowledge or consent and uploading it to her Snapchat account to poke fun at the elderly lady's body. Related to this fear of public exposure and humiliation (or worse) is the practice of many public swimming pools in Germany introduced during summer 2017: they started to offer colourful and highly visible stickers to those customers who wanted to use their smart phones on the premises. These stickers were meant to cover the lenses of the smart phones' cameras in order to avoid suspicion that their users would try to snap some lewd pictures of scantily-clad women or underage boys and girls (Schleutker-Franke 2017). But still: when it comes to the use of our faces for authentication purposes, we tend to find this far more palatable than submitting our fingerprints, as I mentioned above. And, of course, if we fall foul of the law, pictures of our faces will be taken as well in the shape of the (in) famous 'mug shot.' This routine collection of all sorts of facial pictures also implies that there are bound to be vast databases containing these pictures together with other personal information, official as well as unofficial, also as already discussed.

Seen from this perspective, David Lyon's assertion that terrorists "do not pose for photos" (Lyon 2004, 304) is not entirely convincing: if it comes to formerly law-abiding citizens (minor misdemeanours notwithstanding) who (self-) radicalize one way or the other to then try to carry out terrorist attacks such as the perpetrators of the Paris, Brussels, Nice, Berlin, Manchester, Barcelona or (various) London attacks, it can be assumed that pictures for the purpose of passports, ID cards, driver licenses etc., or even mug shots, have been taken at some time. After all, nobody is born a terrorist. Thus, facial recognition software theoretically could pick them up from a crowd. Lyon however is certainly right to argue that terrorists "are likely to use evasive techniques and disguises", and he is also quite right when he cautions that "even if one had some good images, the so-called base rate fallacy means that the chances of false alarms would be very high indeed" (ibid; also see Flaherty 2015). He deems it far more likely that petty criminals will end up being picked up and identified by CCTV systems using facial recognition technology as they, or rather, their faces, are already in a database (ibid.). He provides a telling example for that: "An ordinary crowd of Superbowl fans in Tampa, Florida was scanned using Visage equipment in January 2001 [...] and of the 100,000 about nineteen petty criminals were recognised" (Lyon 2004, 303) – not bad for policing and law enforcement purposes, but admittedly irrelevant with regard to anti-terrorism, and for the purpose of protecting us and our cities from further attacks. Pretty much the same can be said about the Offensive Behaviour at Football and Threatening Communications (Scotland) Act of 2012. This act banned the chanting of certain songs due to their racist and sectarian contents. Over the years, CCTV systems in football stadiums have been used to identify football fans taking part in chanting these banned songs – with dire consequences for all those who could be identified: "Young people have lost out on jobs and promotions because of prosecution, far less conviction; young professionals including trainee teachers and health service workers have faced the prospect of losing entire careers before they started" (Findlay 2017). In Fall 2017, the act was finally repealed.

Regarding the issue of 'picking up individuals from a crowd', it should also be noted that the current limitations of facial recognition software still result in a number of false positives. One of the most recent examples in that regard is the London Metropolitan Police deployment of facial recognition software to scan the faces of the Notting Hill Carnival revellers in 2017, in the hope to detect crime, and to keep the participants of this mass event safe – both of which are entirely laudable reasons. As the London Metropolitan Police explained well before the event, the "technology involves the use of overt cameras which scan the faces of those passing by and flag up potential matches against a database of custody images. The database will be populated with images of individuals who are forbidden from attending carnival, as well as individuals wanted by police" (as quoted in Dodd 2017). The decision alone resulted in strong protests from civil rights groups such as Liberty, Privacy International, StopWatch and Black Lives Matter (BBC 2017a, b, c). But worse was to come: even though the database used on that occasion only contained 500 images, the facial recognition software proved to be faulty – with the result of about 35 false positives, "leading to one incorrect arrest and five interventions" (Lamb 2017). In a 'post mortem' press release, the (independent) Biometrics Commissioner of the British Government acknowledged the failures, admitting that Facial Recognition Technology "has the potential to be a really useful crime fighting tool but we are not there yet. It needs to be properly tested and evaluated if it is going to be effective and it will need to be handled carefully by the police and the government if it is going to be trusted by the public" (Wiles 2017).

As regards the London Met's admittance that 'we are not there yet', when I stated that facial recognition software paired with a high-resolution CCTV system could 'theoretically' pick individuals up from a crowd and identify them, I was hinting at the fact that it is arguably quite a daunting task for facial recognition software to compare a probably somewhat blurred and/or only partial face picture from someone moving through a crowd to compare it with a database of existing frontal face images without creating too many false positives or false negatives. This is a point also made by Introna and Wood who, on the basis of some examples, argue that "locating a face image in the field of vision [of the CCTV system] is not a trivial matter," and that "the effectiveness of the whole system is dependent on the quality of the captured face image" (Introna and Wood 2004, 185, 189–190). Nevertheless, seen purely from a security perspective, this issue as well as all others that I mentioned should not stop us from trying – and neither should the potential of smart CCTV systems be underestimated: if these systems can be made to work reliably and without creating too many false positives or, far worse from a security perspective, false negatives, then they will be able to spot not only faces in a crowd and compare them with a database, but basically any behaviour that is (depending on the location) somewhat out of place: smart CCTV systems could detect suspicious objects, such as abandoned backpacks, flag up intrusion attempts or incidents of loitering that might be further investigated (again, depending on the location), or any other misbehaviour such as theft or vandalism. The detection of any behaviour out of the normal, or the identification of individuals not supposed to be at a certain location, or who are on a watch list, would then raise an alarm that would enable security staff to converge on the potential threat, either via CCTV or in person if deemed necessary. Obviously, this would make it possible to nip at least some planned acts of terrorism in the bud, or to at least mitigate its consequences by way of a swift intervention.

Here, an incident in Brussels' Gare du Midi (Central Station) on 20 June 2017 comes to mind: although it was the alertness of soldiers deployed there and not smart CCTV, their immediate counter action forced the terrorist to detonate his trolley full of explosives prematurely and to flee, to be shot dead in the attempt. Since neither soldiers or police forces or private security can be omnipresent, smart CCTV systems could fill this gap by automatically spotting suspicious behaviour such as a person lingering at one spot suspiciously often (because that person could be a pickpocket), a person running or falling (because that could be due to a safety- or security-related incident), or a person moving against the general flow of pedestrians (because this behaviour is odd), to then alert human operators, who could then initiate appropriate counter actions (Strittmatter 2018). A current example for such a smart CCTV system about to be deployed in a city would be the *Mannheimer Weg 2.0* (roughly translatable as 'Mannheim Path 2.0') of the southwestern German city of Mannheim. It involves 71 CCTV systems deployed at 28 locations and sending live feeds to the control centre of the police. An algorithm developed by the Fraunhofer Institute Karlsruhe scans the stream for untypical behaviour. If such behaviour is detected, a human operator is alerted who will then examine the feed in order to decide whether police intervention is required or not. If so, the planners of the system expect that a police patrol should reach the location in question within two minutes. Of additional interest here is that the algorithm comes with the built-in capability to learn on the basis of image feeds collected which behaviour to ignore and which to flag up. Also of interest is the fact that, mainly due to civil liberties issues, the *Mannheimer Weg 2.0* does not include facial recognition or audio recording, and is pre-set to delete image feeds within 72 hours (Spiegel Online 2018).

A current example of a key asset that already deploys such a system would be the new *One World Trade Center* in New York: here, about 400 smart CCTV systems featuring video-analytical software tailor-made to detect potential threats are already deployed on the premises. Regarding the false positives or negatives flagged up several times by now, I should hasten to mention that this issue seems to be nearly solved by now: the Notting Hill Carnival 2017 disaster notwithstanding, current facial recognition software is more accurate than the systems Introna and Wood commented on about fifteen years ago in 2004 – fifteen years being a very long time in computer science. As Chakya noted five years ago (also a long time regarding computer and software development), the face recognition company Cognitech "boasts a match accuracy rate of 98.75 per cent, an increase of more than 20 per cent in the past decade [while] Facebook recently achieved 97.25 per cent accuracy after acquiring biometrics company Face.com in 2012" (Chakya 2014). Again, it seems that the technical problems of facial recognition can be solved – which 'only' leaves us with the certainly non-trivial privacy issues as seen from the side of civil liberties.

When it comes to sound detection and audio recording, this could also be of eminent interest for smart CCTV systems even though it might also infringe civil liberties – which is why Mannheim opted not to include it as well (apart from facial recognition). In some US American cities such as Los Angeles for example, a system known as *ShotSpotter* has been deployed to automatically detect, triangulate/locate and report gun shots. The more or less precise location is depicted on a map in the control room, enabling the controller to dispatch a police patrol to investigate (Weller 2017). The system is said to be "much faster" than dialling the emergency number (ibid.), and hence has the potential to shorten the reaction time – a time denied to criminals as well as terrorists that could otherwise be used to kill people. So far, the system consists of a network of microphones only – matching such sound-activated microphones with a Smart CCTV solution however is the next logical step forward. And since the microphones are activated by gunshots (or gunshot-sounding noises) only, the impact on civil liberties would be minimal.

8.3 Augmented Reality: Towards *Minority Report*-Style Policing

Since I am already talking about civil liberties issues, this is a good moment to introduce yet another new technology that currently looks very promising from a policing point of view – including a counter-terrorism policing one: *Augmented*

Reality or AR for short. First of all, it is important to note the difference between AR and VR, i.e. *Virtual Reality* which at the moment seems to be rolled out with a vengeance by various computer gaming firms, and to a lesser degree by firms catering for scientific usages, such as archaeology for example. To start with VR, this simulates a more or less realistic alternative reality or environment either via rooms fitted out with large screens or via headsets featuring a small screen mimicking a somewhat restricted 'normal' view. Depending on the system used, and the computing power of the PC creating this virtual reality, the user either is restricted to simply look around, or can, to a certain extent, interact with objects (or simulated persons or creatures) via game controllers. VR headsets such as Oculus Rift, Kinect for Xbox One or PlayStation VR are becoming ever more popular for computer gaming purposes, while the so far most popular game using AR was *Pokémon Go* – a game that was based on catching virtual creatures overlaid into the player's reality. Augmented Reality or AR on the other hand simply 'augments', that is 'adds to', our view of the 'real' world by overlaying it with additional information that might be useful, depending on where we are and what we are looking for. I already mentioned smart apps for our smartphones that can make recommendations for restaurants, bars, shopping malls and suchlike based on our location by overlaying this information with our 'real' view of the world, using the camera lens(es) and the screens of our smart devices for that purpose as an input-output system – these are good examples of AR. Another stunning example (in my opinion at least) is the *Flightradar 24* app: whenever I see the contrails of an airliner high above me in the sky, all I need is to point the camera of my smartphone at it, and the camera image that I see on the screen is overlaid with a box that contains the data of the flight: it's origin, its destination, the type of the airliner, the flight time, and more.

Macmillan's buzzword dictionary defines AR as "the technology of combining real world images, video, etc. with computer-generated information and/or imagery" (Macmillan Buzzword Dictionary). In case that still sounds a bit arcane, Macmillan offers the following explanation that basically sums up what I said above – while adding one crucial new detail:

> Imagine a device which, when you use it to look at something, displays information on whatever you're pointing it at. Focus on a restaurant, and a display pops up giving a menu and customer reviews, look at a historic building, and you're told when and by whom it was built, or point towards that chap sitting opposite you on the metro, and you find out this stranger's name and age. These are the sort of possibilities afforded by the new technology of augmented reality. (Macmillan Buzzword Dictionary; emphasis removed).

Like most of the currently popular VR systems, AR systems basically consist of a wearable computer, an app and a 'human-computer interface' such as the screen of our smart phone that needs to be pointed at the source of our interest – the sky in the case of identifying an airliner, a restaurant in the case of getting information about its menu and its Trip Advisor ratings, or at parts of the street in the case of the aforementioned *Pokémon Go* game. Much more comfortable is a human-computer interface in the shape of a head-mounted display equipped with some sensors (Cowper and Buerger 2003, 19) because then, all we need to do to get AR-based information is to look at things that catch our interest. Of the current AR systems, the probably best-known still is the first one to be introduced as early as April 2013 when it became available to beta testers: Google Glass. It did all that highlighted above, and much more – for example taking 5 Megapixel pictures and (short) videos via voice commands or its touchpad or the MyGlass phone app. As Chloe Albanesius described it one year before the actual release of this gadget, Google Glass "puts your smartphone into a pair of slim glasses and projects its contents for some futuristic, voice-activated fun" (Albanesius 2012).

Unfortunately, and even though the concept itself is quite persuasive since it does away with the necessity to look down on one's smartphone in order to look up information (or hold it in front of one's eyes), the reception was by far not as positive as Google probably expected. This was mainly due to privacy issues revolving around the smart glasses' ability to take pictures and even videos while also recording conversations without anybody noticing or consenting to it. For this reason, a number of facilities such as bars, restaurants and even some casinos in Las Vegas (Clark 2013) banned the use of Google Glasses on their premises even before it was released to the general public in 2014, and a number of lawsuits were launched. Google initially reacted to this unexpectedly lukewarm and even outright hostile reception by stopping the production of Google Glasses in January 2015, but then relaunched an upgraded version of its smart glasses named *Google Glass Enterprise Edition* in July 2017, albeit in a rather modest scale and basically for factories such as Boeing, General Electrics (GE) and DHL only, and not (yet) for the general public (Savov 2017). In any case, Google Glass no longer is the only smart glass on the market – other firms also entered this potentially lucrative market segment in the meantime as well.

But let us return to quite an important detail in the Macmillan Buzzword Dictionary's description of AR devices – the one that mentions pointing the device 'towards that chap sitting opposite you on the metro, and you find out this stranger's name and age.' Not surprisingly, this possibility was yet another privacy concern associated with Google Glass in 2013: that it could, supported by facial recognition software, identify an otherwise unknown person against the person's wish, and even without that person's knowledge (Miller 2013). In the absence of suitable applications ('app' for short), this concern proved to be a bit too far-fetched – or maybe just a little bit ahead of its time: since AR applications quickly gained wide acceptability among at least the more technically minded of the general public, it is also quite probable that AR-equipped devices, including smart glasses, will be incorporated in police work as well in the near future. One force that already makes use of Augmented Reality, albeit only for forensic purposes for the time being, is the Dutch Police. Journalist Timothy Revell describes the most likely scenario as follows:

> You're the first police officer to arrive at the scene: a suspected ecstasy lab. There's drug paraphernalia everywhere, but which piece of evidence could be most helpful for your investigation? Then, a massive virtual arrow appears, pointing out a bottle of chemicals, accompanied by a note saying: 'Bag this please'. (Revell 2016)

As Revell further explains, the officer arriving at the scene uses his or her body cam to live-stream a video of what he/she does and sees to various experts who can then guide the officer either via the officer's smart phone held in front of him/her or, preferably, via the officer's smart glasses (ibid.). This could also be used to alert the officer of something that poses a risk, such as toxic substances or to contamination. As the innovation adviser of the Dutch National Police, Nick Koeman remarked, "[the] technology makes it possible to get the right information to the right people at the right time, in a way that's easy to see" (as quoted in Revell 2016). The 'right people at the right time' could also include the judge and the jury, since AR also makes it possible to capture and map a crime scene in order to recreate it later in 3D, as Cowper and Buerger (2003, 5) already suggested more than a decade ago. Currently, Bedfordshire Police in the UK is trialling such a system, based on Microsoft HoloLens and a mapping software from software company Black Marble (Charlton 2017).

Indeed, AR might well be a "crime-fighting game changer", as Alistair Charlton opines (ibid.). It could also be a game changer when it comes to the training of police officers, or first responders in general. Using both Augmented Reality and Virtual Reality (AVR for short) to "train and assess Security Critical Agents (SCA), including police, fire, emergency medical services, counter terrorism units, border guards and critical infrastructure operators" (LIST 2017) is the aim and objective of a European Union Horizon 2020 research programme named TARGET that commenced in May 2015 and ended in October 2018. The Luxembourg Institute of Science and Technology (LIST) describes the programme that also uses Microsoft's HoloLens as follows:

> TARGET seamlessly blends different approaches allowing AVR systems to be connected to geo-spatial and assessment tools and it also incorporates the use of 3D models captured from advanced photogrammetry drones. TARGET uses state-of-the-art technologies [...] and instrumented real objects to create new mixed reality worlds for use in high stress security training situations. The aim is to improve training while also making it more customisable, cheaper and safer. (LIST 2017)

Again, it is evident that AR, and VR as well, may well be a game changer with regard to policing in general, and counter-terrorism in particular (for more information, see Cowper and Buerger 2003, 33–37; Blount 2010). However, using AR as a tool for training purposes, or for forensics, and thus for investigating and hopefully solving a crime after it had been carried out reminds us of the criticism directed at the early CCTV systems: it definitely is a step in the right direction, but being able to use this technology to prevent a crime in the first place instead of solving it post-hoc would be much better. So, how about equipping smart glasses with facial recognition software as was feared in 2013 when the Google Glass was released? This would require capable smart glasses on the one hand, and a more advanced facial recognition software on the other capable of 'augmenting' the officer's gaze with information of the person(s) he or she is looking at, such as name, residence, occupation, and any convictions or outstanding warrants for arrest or any other relevant information. And of course, it would also either require a certain disregard for civil liberties and human rights as well or a patient and probably long-drawn out campaign to gain the populations' acceptance of the use of AR and VR for police purposes since this would massively intrude into the privacy of perfectly normal and innocent citizens. Maybe that is one reason why such AR-equipped smart glasses have not yet been adopted by the police forces of Western liberal democracies – those states I am focussing on in this book. Outside of the West however, police forces are not so squeamish: in the People's Republic of China for example, such devices, indeed equipped with state-of-the-art facial recognition software, are already in the process of being trialled by various police forces, for example in the city of Zhengzhou. Jon Russell explains the way they work as follows:

> The smart specs [that] look a lot like Google Glass [connect] to a feed which taps into China's state database to root out potential criminals via facial recognition. Officers can identify suspects in a crowd by snapping their photo and matching it to the database. (Russell 2018)

The effortless identification of suspects just by looking at people while being on patrol thus no longer belongs to the realm of science fiction and movies such as *Minority Report* but to our reality – at the moment at least in China, and with the just mentioned limitations: a photo has to be taken first, to be then matched with a database. But since this technology is so eminently (potentially) useful for crime prevention, it is more than likely that it will be improved further, and that it will be soon trialled by a police force near you as well.

8.4 Smart Security: The (Near) Future

When discussing Smart CCTV, we already touched upon a fusion of different types of security technologies – in that particular case, a merger of advanced computer algorithms with high-resolution camera systems capable of detecting any behaviour that is deemed to be (according to the programmed criteria) out of the normal. If we could merge these smart CCTV systems with the advanced ETDs just discussed, then we would be a huge step further towards much improved (though not total) security. In a sense, what we are looking for at the moment is a user- and operator-friendly 'one stop shop' that detects a wide spectrum of potential threats and is both reliable and fast – or, in the words of Singh and Singh (2002, 33), for an "ideal portal [that] would be able to screen a person for weapons, drugs or explosives in real time without impeding the traffic flow." Here, a scene from the Hollywood science fiction blockbuster *Total Recall* (the original one with Arnold Schwarzenegger) comes to mind: the protagonist casually walks through the access tunnel of an underground station while its security system deployed in the tunnel's walls automatically scans him for suspicious substances or weapons that would be highlighted on the greyscale silhouette of his naked body. Unfortunately, at the moment such a system, nicknamed by some experts the 'tunnel of truth', still is "a long way off" (Sheridan 2010) and pretty much in the realm of science fiction – but at least a near-future one, not a distant future fantasy: research in this direction is already making (slow) progress. Or, as Cowper and Buerger (2003, 8) astutely argue, "[yesterday's] science fiction is rapidly becoming a very real part of everyday life."

As I already mentioned several times regarding different technologies, currently we are faced by the fact that most systems are a) not very convenient for those who have to use them, and b) not very reliable either. Again, all we need to do is to think of our experiences at a fair-sized airport's check-in during the busiest time of the day: there are massive queues, and simply too many people have to double back after having been flagged up by the metal detector to put their shoes on the belt as well, leading to further delays: although 'step-on' shoe scanners currently are under development, these solutions still are years off from being deployed – if they turn out to work as hoped for, that is. That this does not sit particularly well with travellers (some of whom actually miss their flights due to unexpected long delays at security) did not escape the attention of officialdom. For example, former director general of IATA, Tony Tyler, admitted that "[managing] security will become increasingly challenging as more of us travel each year. Long queues, intrusive checks and inconsistent procedures create natural frustrations" (as quoted in Calder 2016). Hence, all those who have ever missed a flight due to this major inconvenience as well will probably be very pleased to learn that efforts are being made to address both problem areas, for example by looking to scanning technologies currently used in advanced medicine – especially cutting edge computed tomography (CT) and X-ray diffraction (XRD). The advantage of CT and XRD adapted to security needs would be that there would be no need any longer to put liquids of 100 ml or less (anything bigger than that currently not being allowed on board of planes) in a transparent plastic bag as these technologies would be able to detect and scan them inside the passenger's carry-on luggage. As Tyler hopes, CT and XRD technologies would also ensure 'a slicker screening': "TSA [US Transport Security Administration] is interested in exploring the capabilities of CT-based solutions … in the checkpoint environment to screen for explosives, weapons, and contraband at a significant unit cost reduction" (again as quoted in Calder 2016). However, don't hold your breath, as the saying goes: as Calder cautions, the expected roll-out date of 2020 for such technologies is unlikely to be met – and then, the new technologies will probably first have to be put through their paces at public venues hosting mass events such as sport stadiums or concert halls to weed out teething problems before finally being deployed (Calder 2016).

Much further down the line are systems intended to do exactly what I just described in the Total Recall scenario – systems meant to quickly scan individuals for a variety of suspicious or hazardous materials while also establishing their identity at the same time. One such solution is the Thales Smart Corridor which is designed to make use of 'multimodal' biometrics, that is different types of biometrics and scanning technologies, for that purpose. As Boussadia (iAirport Solutions manager for Security Solutions and Services at Thales) explains, the Smart Corridor "integrates multi-modal identification, detection and inspection technologies including intelligent video surveillance of abnormal behaviour, biometrics and sensor technology including trace, metal and hidden objects and epidemiological risks into an automated screening system" (Boussadia 2009, 8). This integrated corridor system, a prototype of which was shown to the (professional) public in late 2008, is about four to five metres in length. The security devices are not visible to those passing through, and unlike in the case of the Total Recall scene, there is no detailed silhouette displayed to the operators (ibid.). The Smart Corridor was meant to be tested at an undisclosed airport in the first half of 2009, but the fact that no other publicly available sources on this system emerged about ten years after the prototype was introduced hints at the probably rather formidable problems posed by integrating detection and scanning systems (which in themselves are far from absolutely reliable) into a seamless security system. It is also quite telling that apart from Thales, no other private sector company came forward in the meantime to introduce a similar system. So, obviously, for the time being, a corridor-like multimodal scanning, detecting and identification system that not only works under laboratory conditions but also under 'real-life' conditions at busy airports, railway stations and other parts of our critical infrastructure firmly remains in the realm of science fiction. Having said that, such a holistic smart security approach is exactly one of those 'silver bullets' that we are looking for – and sooner or later, we will get there. But we are certainly not there yet.

Chapter 9
Protection: Defensible Spaces

Abstract In this chapter, I discuss passive and mostly 'low-tech' defences in the shape of barricades (i.e. counter-intrusion devices) and citadels (i.e. facilities hardened against terrorist or criminal attacks). After all, in my opinion, these modern avatars of time-honoured brick-and-mortar curtain walls still have a formidable role to play in the times of global terrorism: basically, they are our last line of defence against terrorists who have managed to evade all other high-tech measures described above without being detected and are now ready to strike. Hence, I will take a look at modern city walls such as the City of London's 'Ring of Steel'/Ring of Glass', at modern citadels such as *One World Trade Center* in New York, and at barricades (temporarily) deployed to deny access to certain areas.

Keywords City wall · Ring of steel · Ring of glass · Ring of plastic · Citadels · Residential communities · Barricades · Bollards · Public Space

In the preamble to this section, I already mentioned that I am not only looking at silicon-based high-tech security solutions, and that I would also include passive defences in the shape of barricades (i.e. counter-intrusion devices) and citadels (i.e. facilities hardened against terrorist or criminal attacks). After all, these modern avatars of time-honoured brick-and-mortar curtain walls still have a formidable role to play in the times of global terrorism: basically, they are our last line of defence against terrorists who have managed to evade all other high-tech measures described above without being detected and are now ready to strike.

Again, both the keen interest in, and the sheer necessity of, monitoring people and if need be restricting their movements within one's the city or at least near certain sensitive locations in it did not come into being as a result of 9/11 – rather, it is as old as cities themselves since it is part and parcel of population control. For example, Lewis Mumford argues that in the case of ancient Mesopotamian city states, their rulers applied what he calls "a policy of terrorization" to keep their city-dwellers under control: "[The] necessary co-operations of urban living require the constant application of the police power, and the city becomes a kind of prison whose inhabitants are under constant surveillance: a state not merely symbolized but effectively perpetuated by the town wall and its barred gates" (Mumford 1989,

P. Lehr, *Counter-Terrorism Technologies*, Advanced Sciences and Technologies for Security Applications, https://doi.org/10.1007/978-3-319-90924-0_9

83). It can thus be argued that from the very beginnings, city walls did not only have the purpose of keeping potentially hostile, alien and thus untrustworthy people out, but also to keep potentially restive and quarrelsome people inside, and under firm control.

Modern-day barricades and citadels are no longer just made of brick and mortars, or concrete and steel, but are usually supported by high-tech security solutions such as smart CCTV, which render these variants of walls even more formidable. Here, the City of London's famous *Ring of Glass* is a case in point, and thus will be discussed below. The advantage of this 'wall-building' approach is that within the walled part, life goes on pretty much undisturbed as before, give and take the odd quarrels between neighbours revolving around matters of maintenance, length of grass, or noisy kids. The disadvantage however is that protecting a vulnerable city with such a modernized version of the ancient city wall is only practicable when we are talking about a comparatively small area such as the City of London or the inner city of Belfast during the time of the 'Troubles', for that matter. This also implies that there are, obviously, large areas that lay outside of this wall while still potentially being critical, vulnerable and under threat. Furthermore, building such walls is a risky strategy since if and when this formidable barrier is breached, then we are thrown back to a mere re-action to the terrorists' actions, probably guided by smart CCTV. For that reason, some CIP scholars denounce this approach as a modern version of the ill-advised 'Maginot Line' thinking, and for good reasons: once the attackers have breached the wall either by brute force or by cunning, there is nothing much left that could stop them.

To ameliorate and mitigate the problem posed by city walls as linear and brittle defences, a new urban architecture offers additional, oftentimes 'low-tech' security solutions that can be merged or fused with the silicon-based 'high-tech' ones. Some of these measures are either completely invisible or hardly noticeable at least to the casual observer. Others are potentially more intrusive in a way that, if adopted, they may well lead to a change in the 'feel' of cities protected by them, especially when it comes to open spaces such as parks, squares of boulevards. In this chapter, I will critically examine these solutions, beginning with the least intrusive ones that are basically invisible and geared to protect particular buildings. This will be followed by security measures which easily blend in and, thus, are only noticeable for the well-trained eye. Finally, I will focus on the most intrusive measures that are on offer from the new urban architecture – measures that will change the urban geography of our cities as we know it. And since this chapter is obviously not only about technology as such but also at least touching upon matters of architecture and design as far as it relates to security and counter-terrorism, it is worth repeating Hunt and Kellerman's argument I referred to when I was discussing the basics of critical infrastructure protection: "Vulnerabilities are the only element of risk over which we have some control. Vulnerabilities are built into a system through its design, the equipment that is used, the procedures that are followed, and the policies that are made" (Hunt and Kellerman 2000, 3). This means that such vulnerabilities, if detected early enough, can be 'designed out' from the first blueprint stage. And this process of 'designing out' as a very important element in our defensive measures against terrorism will feature in this chapter as well. But let us start with the basics, and thus with city walls first.

9.1 City Walls: The Re-Emergence of an Age-Old Concept

When it comes to defending our cities against an external threat, a time-honoured way to do so is to protect the town or city in question with a wall, usually combined with other, supporting defensive features. Examples would be moats to prevent the risk of undermining and collapsing sections of the wall (not so much to act as a water obstacle), labyrinthine narrow passageways within the walls to render the enemy's advance more difficult, walled family or clan compounds only accessible through one narrow door or a ladder, sudden cul-de-sacs, clockwise-winding staircases with uneven steps aiming at the invader losing his (rarely her) step while restricting the movements of his (usually right) sword arm. As already explained in the first part when I discussed the 'new military urbanism', defences were not necessarily aimed solely at an external enemy such as hostile levies besieging the city – quite often, the defences within the walls were aimed at potentially hostile neighbours: either those of a different ethnic or ethno-religious make-up, or those of a different class. In the case of the former, deeply untrusted Chinese migrant minorities usually were kept well under the guns of certain Spanish and Dutch colonial cities in Southeast Asia (such as Manila and Batavia/Jakarta), to also act as a wave breaker in case the Filipino or Malay majority would try to oust the unloved colonial overlords; in the case of the latter, the ruling elites routinely built citadels as a place of last resort, both as a security measure against the external enemy and the internal enemy. Oftentimes, potential internal enemies also consisted of rival feudal families competing for control. An example for that would be a number medieval Italian city states such as Florence and San Gimignano that feature fortified clan towers into which beleaguered feudal families could withdraw. In more modern times, Baron Haussmann's ambitious rebuilding of Paris would be another example: as already explained, it did not only aim at improving the rather squalid living conditions and the poor sanitation: the new boulevards and squares also aimed at keeping the citizenry under control. As he himself readily admitted, his renovation of Paris was mainly about 'ridding old Paris of the quarter of riots and of the barricades' (as quoted in Moncan 2012, 34, translation from French mine).

The overall aim behind all these defensive measures was to create what Oscar Newman calls 'defensible space.' For Newman, defensible space "is a surrogate term for the range of mechanisms – real and symbolic barriers, strongly defined areas of influence, and improved opportunities for surveillance – that combine to bring an environment under the control of its residents" (Newman 1972, 3). In our present context, Creekmore's definition is more to the point: "Defensible space is a conscious effort by residents and city planners to define urban space by erecting barriers or markers that indicate ownership, enhance security, and discourage or manage passage" (Creekmore 2014, 51–52). The City of London, that is London's financial district, is an interesting and quite instructive example in that regard. Attempts to fortify and secure the City of London precede the current threat posed by Al Qaeda and ISIS by a decade: these attempts commenced when the Provisional Irish Republican Army (P-IRA) started to launch attacks on the British mainland in 1990 and were intensified under the impact of the Bishopsgate bombing attack in April 1993. In the aftermath of this attack, some not further specified 'leading City figures' suggested far-reaching changes to the City of London, amongst them the demand that "[the] City should be turned into a medieval-style walled enclave to prevent terrorist attacks [...] with access through a number of small 'gates' and controlled by security discs (Sivell 1993, as quoted in Coaffee 2003, 106–107). Otherwise, the financial institutions so far present in the City of London might have to reluctantly consider moving to safer places out of the reach of the P-IRA – places such as Frankfurt or Zurich... These 'leading city figures' quickly got their way by way of what became colloquially known as the *Ring of Steel* – a modern city wall that was based on an earlier one in the shape of the Ring of Steel protecting the inner city of Belfast, also against P-IRA attacks. Like its earlier Northern Irish avatar, the City's new defensive perimeter consisted of a network of high-tech and low-tech protective measures, such as private and public CCTV systems regarding the former, and concrete or steel roadblocks plus checkpoints manned by armed police regarding the latter. Furthermore, highly visible police patrols were deployed as an additional deterrence – not only against P-IRA terrorism but also against 'normal' crime. The idea behind it was the same as in the earliest variants of city walls: nobody should be able to enter unobserved and unchecked. On top of that, and heeding the lessons of the Maginot Line, individual buildings deemed to be at risk were gradually turned into veritable fortresses or 'citadels' in their own right, also protected by access controls, and bristling with audio-video CCTV systems which hardly left any blind spots.

According to Coaffee (2004, 278–293), the City of London's Ring of Steel went through several phases, depending on the waxing and waning threat posed by the P-IRA. In the first phase ('apprehension') from the beginning of the P-IRA's mainland attacks during 1990 to the bombing of St Mary Axe in April 1992, city-wide security measures were under discussion, while a routine of overt police patrols was introduced as a stop-gap measure. The second phase ('containment') that began as a result of the St Mary Axe bombing and the Colman Street bombing of June 1992 and ended with the Bishopsgate bombing of April 1993 introduced armed police checkpoints, an enhanced traffic management system and CCTV systems. The Bishopsgate bombing of April 1993 inaugurated the third phase ('deterrence') that lasted until September 1994 when the P-IRA declared a ceasefire. It saw the construction of further checkpoints on the one hand, and the deployment of advanced public and private CCTV systems on the other. The fourth phase ('optimism') lasted from the P-IRA declaration of a ceasefire in September 1994 to February 1996 when the Docklands bomb demonstrated that the ceasefire was over. During this optimistic phase, visible police presence was reduced, while previously improvised police checkpoints were made permanent. Also, a further upgraded police-operated CCTV system was fielded.

Under the impression of the Docklands bombing, a one-year long fifth phase began ('reactivation') during which visible police presence was increased again, while a number of 'roving' checkpoints were added to the permanent checkpoints as further security measures. On the legal side of things, the police were given increased powers regarding the 'stop and search' of suspects. In a sixth phase between February 1997 and June 1999 ('extension'), the City of London's Ring of Steel was expanded, and additional advanced CCTV systems were deployed. As a consequence of the Good Friday Agreement of 10 April 1998 between the British government and the P-IRA, the seventh phase from June 1999 to September 2001 ('reappropriation') adapted the security measures taken against the P-IRA to a new threat posed by anti-capitalism protesters, while also keeping dissident Irish republican terrorists in mind. In this phase, building security was enhanced, and a better rapport between the City of London police and other forces established. In an eighth and (so far) final phase ('reappraisal') as a reaction to 9/11, counter-terrorism measures were re-examined, and alert systems were improved. Of special interest in this context is the increasing importance of advanced CCTV systems operated by the police and private parties, which gradually turned the *Ring of Steel* into a *Ring of Glass.*

The *Ring of Steel* had a couple of disadvantages. First of all, as a *Times* article pointed out, it actually turned the City of London into an even more tempting target for the P-IRA "because of the propaganda value to be derived from penetrating that loudly trumpeted ring" (O'Brien, as quoted in Dillon 1996, 292–293; also quoted in Coaffee 2004, 288). Penetrating this ring was by no means impossible, as for example the Docklands bombing demonstrated: due to its dimension, parts of the vaunted ring only consisted of rows of traffic cones for temporary road management and access restrictions leading to an improvised check point – a fact that did not escape the attention of many observers, some of whom quipped that the 'Ring of Steel' in reality was more like a 'Ring of Plastic' (Coaffee 2003, 100, 114). Secondly, as the *Times* article also highlighted, all these measures tied down substantial numbers of police officers who were then not available elsewhere: "The other way in which the charade increases the risk to the city is that it diminishes manpower available to counter the IRA threat. Fixed roadblocks need a lot of trained manpower" (O'Brien, as quoted in Dillon 1996, 292–293; also quoted in Coaffee 2004, 288).

To a certain extent, a 'Ring of Glass' can redress these problems: on the one hand, it is less obtrusive, and might thus be less of a 'red rag' for potential attackers – now in the shape of ISIS-related terrorists. On the other, it certainly saves much of the otherwise required trained manpower (police officers, mostly) since many of the advanced CCTV systems can be monitored by comparatively few operators. As described above, these operators can then guide police response teams to areas in which some unusual activity was discovered. What can be discovered by these advanced CCTV systems obviously depends on their capabilities: CCTV cameras equipped with ANPR technology are quite reliable by now, and of course many of these cameras also allow the operator to zoom in on individuals who seem to behave out of the expected behavioural patterns of the ordinary passers-by. However, this is pretty much a manual process: an automated video-analytical facial identification as discussed above still seems to be beyond the capabilities of most, if not all, of the deployed systems, all grandstanding and industrial promises notwithstanding. Other possibilities to identify individuals such as gait recognition may be in the proverbial 'works', and so may be systems that can sniff explosives from a distance – but again, 'don't hold your breath', as the saying goes.

However, as I argued in the first part, even a swift reaction guided by smart CCTVcould result in the loss of lives of citizens. This also implies that resilience-based approaches that are aiming at the survivability of key assets are not necessarily what we are looking for in our context, or at least not as stand-alone solutions: obviously, it would be of cold comfort for the victims of terrorist attacks to die in the knowledge that the airport, railway station or mall in which they perish would survive, and back in business just a couple of hours after the incident. They would probably argue that it would have been better to stop the terrorists from striking in the first place by detecting them early enough. And this very understandable sentiment brings us to the question of how we can optimize the modern version of the old idea of city walls and integrate some or all of the security systems discussed individually above.

9.2 Citadels: Hardening Key Assets Against Terrorist Attacks

Again, the realization that some buildings are important or 'critical' enough in one way or another to be hardened against a threat is nothing entirely new, and nothing that came to our attention only after 9/11. Basically, it can be argued that as soon as towns and cities came into being around 6500 BCE in Mesopotamia, the necessity of protecting certain structures critical for survival became obvious. Of course, the seats of power at the centre of the town or city probably were the first to be secured – against enemies from outside the city walls but also against enemies in the shape of dissenters and the then not yet so loyal opposition in the shape of other potential strongmen. The urge to fortify the seats of power is the origin of the so-called *citadel*, which literally means 'little city,' as the fortified core of the city. It was protected by its own curtain walls and towers, and accessible only via a few fortified gates with guards being on duty to check who was passing through them. In times of war, it also served as a last line of defence if, usually after an extended siege, the outer walls of the city were breached by the enemy. Temple precincts and their inner sanctuaries also immediately come to mind because of the valuables they usually contained, or granaries because of the bread stored there – which, in hard times, made the difference between starving to death and surviving. By their very nature, these buildings were meant to be easily accessible, and hence more difficult to secure. Thus, finding a balance between accessibility and security was not always easy even in the olden days. Several millennia later for example, in medieval England when warfare between competing kingdoms was rife, coastal villages and towns tended to store their valuables including gold and silver on some islands dotting the coastline, such as Lindisfarne. This made them safe from squabbling kings and princes as well as their marauding soldiers – but eminently vulnerable to another and at that time still unexpected type of predators: the sea-raiding Vikings, in the case of Lindisfarne, which was sacked and looted in the year 793 precisely because of the presence of easily portable (and removable) treasure. It is thus hardly surprising that churches and monasteries were turned into veritable fortresses under the impression of this incident that could be seen as a watershed moment comparable to 'our' 9/11.

In our more settled times, one could be forgiven to think that citadels at the core of our cities are no longer required. However, just as in the case of the reinvented city walls as 'walls of steel' followed by 'walls of glass', citadels still have their utility when it comes to certain types of buildings. The already mentioned headquarters of the Bank of England in Threadneedle Street, London, is such a modern citadel – even though it does not look like one to the casual passers-by. A closer look however would reveal several CCTV systems along the façade of the building as well as other security measures when approaching the entrance. The same can be said for the headquarters of the intelligence services where anybody entering must pass through several security checks before gaining entrance – and then still gets accompanied by staff everywhere they go. And of course, in an era of global terrorism, seats of government also are well secured nowadays and not usually accessible to the public any longer as they might have been before. Interestingly, this is handled differently in the cases of parliament buildings or the Capitol in Washington: since these buildings represent the people, and thus the public, they are at least semi-accessible even when parliament is in session: visitors are welcome in certain areas, while others remain inaccessible to them. Still, checkpoints with metal detectors and suchlike control and monitor the entry of people, and so do CCTV systems inside of these buildings.

Turning individual buildings into modern-day citadels, or 'citadelizing' them to give it the proper term, of course is also possible for other buildings, be they office buildings, hotels, shopping malls, schools or any other individual part of our built-up infrastructure that is deemed to be critical, vulnerable and under threat, which means at risk. This process of citadelization is rather easy in the blueprint stage of a new building in which security features simply can be designed in. The new *One World Trade Center* in New York on the ground of the World Trade Center is a telling case in point here: taking the lessons learned during 9/11 on board, it features a reinforced concrete base, extra-wide pressurized stairwells, and biological/chemical air filters on top of that. Furthermore, the interior is constantly monitored by about 400 smart CCTV systems featuring video-analytical software tailor-made to detect potential threats (Tarantola 2011). Retrofitting already existing buildings with similar security features is, on the other hand, more difficult: smart CCTV systems can of course be added, and so can filter systems. The original panes of the glass fronts also can be replaced by shatterproof glass that, in case of a major terrorist attack or an earthquake for that matter would not rain down in the shape of lethal, dagger-like splinters, thus helping to minimize casualties. The entry points of the building can be fitted out with various kinds of detectors, while bollards, concrete or steel flower pots or artworks made of suitable material can be used as unobtrusive barriers protecting said building from car or truck bombs. Even sturdy trees can serve that purpose. But measures such as reinforced concrete bases or blast-proof stairwells are either not possible at all or prohibitively expensive – which, putting it bluntly, means that most of our skyscrapers are still vulnerable to any reruns of 9/11.

Currently, international airports again are the best examples of on-going citadelization for obvious reasons: carrying out a successful strike against an airliner in mid-flight still is seen by many terrorist groups as the pinnacle of terrorism – which is why Al Qaeda and affiliated groups like Al Qaeda in the Arabian Peninsula (AQAP) invented ever more sophisticated devices in the hope to smuggle them on board and detonate them there: liquid bombs, shoe bombs, and even an underwear bomb. If such audacious plans fail, then the next-best target seems to be the terminal building itself, as the bombing attacks of Moscow Domodedovo (24 January 2011, 37 casualties) and Brussels Zaventem Airport (22 March 2016, 32 casualties) demonstrated. Hence, it makes eminent sense not only to protect airliners from attacks, but terminal buildings, especially the 'landside' part prior to the security checks, as well by 'citadelizing' them – either by retrofitting already existing buildings with available safety and security features, or by including such features in the blueprint stage when the airport is planned.

An example for such a blueprint-stage citadelization comparable (in parts) with the one of One World Trade Center is Rome's Leonardo da Vinci Airport (also known as Fiumicino Airport). When the airport was expanded in the years immediately following 9/11, the need for enhanced security was still prominent on the planners' mind. Hence, a decision was made to group the check-in for those airlines most prone to be attacked, that is all US airlines and the Israeli Airline El Al, together in one new terminal, Terminal 5. Kilcullen comments on the security features as follows:

> [Terminal 5] is set back from the road, and surrounded by barriers to block car bombs, in its own self-contained compound a safe distance from the other terminals. Passengers and bags are screened in a section of the building that has blast barriers disguised as partitions, extra security personnel, fire-resistant furnishings, and a roof system designed to pop off and dissipate the blast without collapsing the building or bringing the ceiling down […]. Chairs and tables are bolted to the bare concrete floor, there's no carpet or drapes – the building is constructed to minimize blast or fire hazard in the event of a bombing. (Kilcullen 2016, 118–119)

As in comparable cases such as the new Terminal 2 of London Heathrow International Airport (LHR), these additional security measures usually remain unnoticed by the passengers as long as they are not too 'in your face' and not too 'brutalist', i.e. interfering with the overall experience of travel. After all, many passengers, including frequent flyers but especially occasional or 'first time' ones, tend to be full of anticipation and, thus, on the nervous side of things. The impression of being corralled into a prison-like holding pens would not do much to help dissipate these feelings. There is one big 'however', however: it needs to be acknowledged that adding security measures as just described does not necessarily eradicate the threat – rather, these security measures fall into the bracket of 'consequence mitigation', that is minimizing the impact of an attack instead of preventing it. Even barriers, bollards, bomb-resistant bins or concrete/steel 'ornamental' flower pots fall into this bracket since they only harden a terminal against a truck or a car bomb – they don't stop a passenger with a trolley full of explosives-filled suitcases from accessing the building. But hardening airport terminals against attacks in a way to render them more resilient and blast-proof is only a first step – adding active security measures such as scanners and CCTV systems is the next one. Again, airports are exemplifying this seamless (well, at least in theory) fusion of brick-and-mortar-based security measures with silicon-based ones.

The fusion of passive security measures such as barricades, bollards, stainless-steel flower pots and blast-resistant bins with more proactive security measures such as Smart CCTV system already starts way before the barrier known as security check that a passenger has to negotiate to proceed from 'landside' (i.e. the less-protected part of the airport before security checks) to 'airside' (the part one accesses after security check and passport control). At the moment, smart CCTV systems that are 'intelligent' enough to alert human operators of a behaviour that violates normal patterns are the state of the art. In the future, more sophisticated sniffer systems will become available that can detect traces of explosives, or of chemical, biological or radiological signatures, hopefully with enough accuracy to detect the source, as discussed. For the foreseeable future, it is however only the 'airside' part of an airport that is fully citadelized. To get there as a passenger, one has to pass security check, which means the familiar array of various scanners and ETD systems, backed up by physical screening and patting-down if and when required. To get there as a member of staff, or an employee of one of the shops, bars and restaurants airside, one has to either get through checkpoints controlled by security staff and very similar to the ones the passengers have to pass, or to access-controlled gates that may or may not make use of biometrics, depending on the airport in question. Once there, one should theoretically be save – as long as no mistakes were made, or as long as the terrorists did not develop yet another way to sneak either themselves or some explosives through this formidable array of security technology. And once there, one could actually stay there: shops, restaurants, cinemas, sleeping pods or airside airport hotels (at least in the larger international airports) basically offer all amenities one needs.

9.3 More Citadels: Hardening Residential Communities

Of course, and despite some anecdotes about people getting stranded airside for weeks or even months for a variety of reasons, airports are not meant for people to strike roots and simply remain there. The same can be said about modern city malls that also tend to get more and more citadelized (at least the more upmarket ones) with access controls and scanner systems. Other examples for citadelization would be office buildings and (again upmarket) apartment blocks or condominiums: once inside, one is save – be it from terrorists, or criminals, or just from panhandlers and so-called 'chuggers' ('charity muggers', that is aggressive street fundraisers), or any other people one would identify as undesirables. Security only is temporary, however: as soon as work is done, or shopping completed, we have to negotiate open and public spaces as potentially dangerous areas on our way home as our personal security bubble – which implies that we have to repeat the same commute between the security bubbles of home, office, and shopping/entertainment centres basically every day. Hence, the next logical step in the process of citadelization is creating a security bubble that merges these smaller 'home, work, play' bubbles into a big one. As Marcuse says, "[the] new form is for citadels within buildings or fortified complexes, including more and more of the facilities necessary for daily life within the building itself. One will never have to leave the citadel for shopping, for meals, for entertainment, for personal services" (Marcuse 2004, 271). Marcuse uses the World Trade Center prior to 9/11 as an example: here, people could arrive with the underground system, spent the whole day in the complex to work, eat and shop, to finally commute home again using the underground, with no need to get out of it to walk the streets of the 'mean city' itself. As a consequence, Marcuse argues, this "pattern, pre-existing the attack on this particular citadel, will be strongly accentuated, but in less high-rise, less representative, less 'signature' fashion, and more heavily barricaded and secured even than before" (Marcuse 2004, 271). Although, probably still under the impression of 9/11, Marcuse probably was a bit too pessimistic as regards the future of high-rise, signature buildings, the core of his argumentation is a sound one: security arguably is highest if one has the choice to remain in a citadel basically all the time if one is inclined to do so.

Concerning Marcuse's mentioning of 'less high-rise, less representative' buildings, another safety and security-related trend needs to be mentioned here: the trend towards residential guarded communities. In its geographic scope, it seems to sit somewhere in the middle between 'city walls' and 'citadels', more concentrated than the former, but usually larger than the latter. That groups of affluent middle to upper class citizens turn their back to the inner cities for fear of crime, and now also of terrorism, is not really a new trend: in the introduction, I quoted Teresa Caldeira (2000, 1) as saying that "[in 1980-2000], in cities as distinct as São Paulo, Los Angeles, Johannesburg, Buenos Aires, Budapest, Mexico City, and Miami, different social groups, especially from the upper classes, have used the fear of violence and crime to justify new techniques of exclusion and their withdrawal from traditional quarters of the cities." And it is also not necessarily a uniform one: the problem many modern cities in the West have is no longer the outflow of affluent citizens, but rather the inflow of them which drives up the price of real estate, thus pushing out the lower classes in a process known as 'gentrification.' How the trend looks like depends on the particular city – but that does not matter in our context. What matters is that these residential communities feature their own protected perimeter, and also their own private security guards in order to enhance the safety and security of its inhabitants simply by excluding everyone who does not possess the required credentials such as those discussed in the chapter on biometrics.

The implications of this process of citadelization even of residential areas on top of malls, government buildings and the like are eloquently illustrated by Flusty. For the case of Los Angeles, he argues that a city undergoing such a pronounced process of citadelization basically changes its character, turning into a "city consisting of numerous fortified cores of private space, each augmented by more permeable outer perimeters of contorted paths, lights, motion detectors and video cameras projecting into the public realm of sidewalks and streets" (Flusty 1994, 37). Another issue worth to be discussed is the proliferation of private security, a business which seems to grow hand in glove with the rise of these modern citadels and city walls. I shall discuss this issue and its implications (which are not always that obvious) for our open societies at length in the third part of the book.

9.4 Barricades: Hardening the Cityscape

Incidents such as the car attack of Dijon on 21 December 2014, the van attack in Nantes just one day later and the refrigerated truck attack in Nice on 14 July 2016 demonstrate that the protection and citadelization of some key facilities such as airports, government of office buildings, malls or sport stadiums and suchlike is not yet enough. Rather, vulnerable open spaces in the shape of parks, shopping roads, beach promenades and comparable venues also need to be protected as well – especially during mass events that are known to draw thousands of spectators, such as the French Bastille Day (14 July), or the popular New Year's Eve celebrations in basically all major cities around the world. This lesson was not lost for example on the organizers of the popular annual Paris-Plages beach festival that opened just six days after the Nice attack of 14 July 2016. Matthieu Lamarre, a spokesperson of the government, explained that "[after] what happened in Nice, we are in the process of setting up devices aimed at blocking vehicles entering the site with barriers, other vehicles standing in the way, and blocks of concrete" (as quoted in Blamont 2016).

The devastating Christmas Market attack half a year later in Berlin, 19 December 2016 showed that, unfortunately, this was not the case for neighbouring Germany. According to security experts interviewed after the incident during which 12 people were killed and 56 injured, the opinion seemed to have prevailed that additional protective measures were not necessary since no major attack had happened in German cities thus far. Like many other experts, Carsten Baeck of the private security provider Deutsche Risikoberatung, criticised the decision not to make more efforts to secure mass events such as Christmas markets – even though these events were known to be potential targets for terrorist attacks. As he points out echoing Lamarre, it is relatively easy to improve the security of such events with simple means, for example by using trucks and cars as improvised barriers (Doll 2016). But, as I already mentioned in the first part of this book, what we do to increase our security tends to be *reactive* instead of *active* – not always, but most of the time. Another example for this would be the installation of security barriers on several bridges in London after the London Bridge terror attack on 3 June 2017: these concrete and metal barriers were "put in place overnight on Waterloo, Westminster and Lambeth bridges [as a] precautionary measure […] after seven people were killed and 49 injured in the deadly van and knife attack on Saturday night" (Forster 2017). Ironically, London Bridge had featured such a barrier before, but it was dismantled due to complaints of cyclists who deemed it to be a trip-hazard. In that particular case, entirely reasonable safety concerns obviously trumped more abstract security concerns – until the attack led to second thoughts on that.

The oftentimes tricky question on how to get the balance between safety and security right finally brings us to at least three flip sides of barricading whole cities. The first one is quite obvious: retrofitting obstacles such as bollards or barriers to certain areas might interfere with the free flow of movement through the city, as already mentioned (see for example Zheng and Hall 2003). Pedestrians and cyclists might have to jostle for space, or cyclists and motorists, depending on the location. The second one also is fairly obvious, at least to long-time city dwellers who can witness the changes brought to 'their' city over time: in the words of Peter Marcuse (2004, 272–273), "'Public space' will become less public; free access and free use will be severely limited. By contrast, controlled spaces, such as malls, will increase their attraction. Some public spaces […] will simply be barred for open use, or so tightly controlled as to inhibit activities normal to a democratic society." The third is that our modern inner cities may well end up as "a fortress of office blocks whose streets empty at dusk", as Kaplan (2000, 85) puts it, while also pointing out the following: "We should also recognize, though, that the architectural reconfiguration of our cities and towns has been an undemocratic event – with decisions in effect handed down from above by an assembly of corporate experts" (Kaplan 2000, 86). We will return to this issue and its implications for society later in the third part.

Chapter 10
Threat Displacement Instead of Threat Eradication: Some Concluding Caveats

Abstract In this brief concluding chapter of the book's second part, I will offer a couple of warnings or 'caveats' before moving on to the more fundamental discussion of the merits and demerits of using technology as a sliver bullet against terrorism in the third part. First of all, I argue that the introduction of available technologies, especially off-the-shelf solutions, has been a rather slow and tedious one. Secondly, I argue that all the security measures discussed do not completely eradicate the threat posed by terrorists: they rather displace it to areas just outside the defence perimeters. And thirdly, I point out that some attacks simply cannot be prevented as long as we are not prepared to deploy plenty of sniffers, scanners and other detection devices all over the city.

Keywords Threat eradication · Threat displacement · 'Landside' attacks · Knife attacks · Function creep

After the description and discussion of current and near-future security technologies already deployed, or soon to be deployed, to counter the threat of terrorism, a couple of warnings or 'caveats' are in order now before we move on to a more fundamental discussion of the merits and demerits of using technology as a sliver bullet against terrorism. First of all, and as a first caveat, it has to be said that the introduction of available technologies, especially off-the-shelf solutions, has been a rather slow and tedious one. Even at our airports as probably the most sought-after terrorist targets, and thus the parts of our critical infrastructures most at risk from a terrorist attack, the progress in terms of introducing available state-of-the-art technology was, and still is, a painfully slow one (Calder 2016). This is even more the case outside of such key assets where additional security measures are deployed only reluctantly and after a major incident, and then only very locally – for example bollards after the London Bridge terrorist attack on 3 June 2017. After all, high-tech security does not come cheap, while low-tech measures such as barricades or bollards tend to interfere with traffic flows, or inconvenience pedestrians and cyclists by further restricting the space allocated to them. Also, the tendency of institutions seen as bureaucracies to be cautious and conservative as regards the introduction of new technologies as this could upset 'business as usual' should not be

P. Lehr, *Counter-Terrorism Technologies*, Advanced Sciences and Technologies for Security Applications, https://doi.org/10.1007/978-3-319-90924-0_10

underestimated – a point forcefully made to me by a former UK Home Office official with some interesting examples which, alas, were of a very sensitive nature and thus cannot be mentioned here.

With regard to the second caveat, and as I already mentioned when discussing barricades, all these security measures do not completely eradicate the threat posed by terrorists: they rather displace it to areas just outside the defence perimeters – in the case of airports loosely defined as an area where passengers and their companions as well as so-called 'meeters and greeters' need to pass through either on their way in or their way out. In aviation parlance, the area outside the security checks is known as 'landside', while the areas after security are called 'airside.' These 'landside' areas, such as departure and arrival, most of the time are full of people who enter and leave without yet having been controlled. And most of them carry heavy bags and suitcases, or even trolleys packed with them. We do not need to speculate what could happen if terrorists were to attack these areas, because such attacks already occurred. In the previous chapter, I already mentioned the attacks on the airports of Vienna and Rome on 27 December 1985, carried out by members of the Abu Nidal Organisation (ANO). In these two coordinated attacks, 16 people died while 120 were injured. ANO's weapons of choice were assault rifles and hand grenades, which is why I categorized these attacks as shooting sprees and acts of conventional high impact (CHI). Nearer to the current time, on 24 January 2011, Moscow's main international gateway, Domodedovo Airport, was targeted by a Chechen suicide bomber who set off his improvised explosive device (IED) in the baggage area of the arrival hall. In this act of suicide terrorism, 37 people died, and a further 173 were injured. And finally, on 22 March 2016, two suicide bombers set off their explosives hidden in suitcases in the check-in area of Brussels Airport, while a third suicide bomber obviously lost his nerves and fled, as already mentioned. 16 people died in these two blasts, while several dozen escaped with injuries.

Now, while even airports suffer from a 'displacement effect' when it comes to improving their security, other parts of the mass rapid transport systems of our times are far more difficult to secure due to the sheer number of passengers passing through them day in, day out. In most countries, these are the main railway stations and metro/subway stations; in some countries, this also includes major ferry terminals. Again, what could happen there if terrorists were to attack them is nothing we need to speculate about in the case of rail and metro systems: these have been regularly targeted by various groups over the last decades. The first such incident goes way back to the end of the nineteenth century when the Irish Fenians attacked the new London Underground: on 30 October 1883, they set off two bombs that injured more than 70 commuters (British Newspaper Archive 2013). In our century, London Underground was targeted again in 2005 by the London 7/7 suicide bombers, and again, just two weeks later, by the Al Qaeda-affiliated London 21/7 bombers – who however fortunately failed to trigger their explosives due to a suboptimal mixture of the explosives. One year prior to this incident, Madrid's Atocha train station was the target of Al Qaeda-affiliated bombers. On 11 March 2004, the perpetrators left several bombs in commuter trains bound for, or just leaving, this station. Ten of them detonated, killing 192 passengers and injuring more than 1000 others. Allegedly, the terrorists also hoped that the glass roof of the station would cave in because of the blasts, killing and maiming even more people. Also in 2004, Chechen suicide bombers wreaked havoc in Moscow's Metro on two occasions – on 6 February 2004 and on 31 August 2004. In those two attacks, 51 people were killed while 170 were injured. It also should not be forgotten that the second target of the Brussels 2016 attacks was a metro station. Here, another 16 people were killed.

Although it would be a logical thing to do, trying to control and monitor the access to these railway and subway stations in a similar way as we do already control access to the airside part of an airport can be ruled out for the time being at least: current metal detectors and scanners simply are not fast and reliable enough to deal with a constant stream of people, many of them quite in a hurry especially during rush hour. One only needs to imagine the queues thronging up at, say, Oxford Circus in London around 5 pm to get the picture. Currently, the only obstacle between them and the platforms are the ticket barriers. The same obstacle exists in major railway stations throughout the United Kingdom, for example in London King's Cross, or at least for most platforms at Edinburgh Waverley Station. Even there, the queues during rush hour can be quite daunting – nobody would want to make them even longer courtesy of metal scanners and the like. Hence, until 'smart corridors; as discussed above finally can be made to work as advertised, we have to take our chances here, and trust the statistics I mentioned in the introduction that demonstrate how unlikely it is to become a victim of a terrorist attack – or hope for a swift reaction by well-trained police in the case that we run out of luck.

There is an uncomfortable truth lurking beneath the surface here in any case: some attacks simply cannot be prevented as long as we are not prepared to deploy plenty of sniffers, scanners and other detection devices all over the city. A telling case in point here comes from the front of youth crime in the United Kingdom: there seems to be a steady rise of so-called 'knife crime' in the shape of stabbing attacks, with "the death toll among under-25s in London [almost doubling] year-on-year" (Dodd 2018). A variety of counter-measures have been discussed and implemented over the years, with only moderate success – among them portal metal scanners like those used at airports at school entrances. ISIS-related terrorists recently have used knifes as well, for example in the case of the attempted beheading of a British Muslim soldier in Birmingham in 2007 (BBC 2008), the murder of Fusilier Lee Rigby on 22 May 2013 in Woolwich by two attackers who ran him over with their car before hacking and stabbing him to death (BBC 2013), or in the case of the London Bridge attack on 3 June 2017 where three attackers stabbed visitors of the Borough Market area, also after having run down several people with their van prior to that (BBC 2017b). While the vehicle attack phase of the latter two incidents could possibly have been prevented by barriers, the knife attack phase of these incidents could only have been thwarted if there had been metal scanners through which the flow of pedestrians had to move. This was not the case, and even if so, the terrorists simply would have chosen other locations. Unfortunately, this threat displacement is unavoidable for the foreseeable future – at least as long as we are not prepared to turn our cities into a conglomeration of citadels, loosely connected to each other with numerous access gates controlled by security personnel in order to monitor the flow of pedestrians and vehicles.

There are indicators that such a development might eventually occur, and I shall discuss that in the next section. This however still is in the future – if it comes about at all. In the here and now, all that can be done is deploying more CCTV systems, smart or not, and employing more security personnel. And this leads us to a third caveat: mission creep, or 'function creep', as others call it. Coaffee flagged that up when discussing the Automated Number Plate Recognition (ANPR) systems, ostentatiously rolled out in London as a counter to traffic congestions. However, for the purpose of countering crime and, especially, terrorism, adding a facial recognition capability to the cameras reading the number plates was (and still is) also mooted – as I mentioned above. Not surprisingly, as Coaffee follows, this did not sit well with civil rights activists. He quotes Gareth Grossman of Liberty as saying that "[there] is an issue we are concerned about which is called 'function creep.' This is where we are told that a system is being set up and used for a certain purpose and then we find out it is being used for another totally different one. It is a dangerous precedent" (as quoted in Coaffee 2004, 293). Grossman certainly is not wrong on that. For example, the new legislation against money laundering pushed through the British Parliament after 9/11 in order to combat the funding of terrorism was first used in anger against certain Icelandic banks during the 2008–2011 Icelandic financial crisis. And the equally new Regulation of Investigatory Powers Act (RIPA) of 2000 that regulates when and how public bodies can carry out investigations and surveillance (including the use of communication data) to combat serious crime was mainly used (or abused) against suspected benefit cheats, dog-walkers suspected of not clearing up behind their dogs, or households suspected to ignore bin rules (Asthana 2016; Wardrop 2011) – while one British police force used it to spy journalists (Greenslade 2014) and another on one of their own in relation to an employment case (Peachey 2016). Obviously, the availability of modern counter-terrorism high tech on the one hand, and the sweeping powers bestowed on authorities that enable them to use these new tools on the other, have consequences for our open societies, and even for our liberal democracies themselves. This observation finally brings us back to the questions I asked in the introduction: how far can we go, and where will we end up? These are the two questions that I will focus on in the next section.

Part III
Brave New Cities: The Law of Unintended Consequences

In the previous part, it already became obvious that applying technical solutions to combat the scourge of terrorism may not necessarily be the 'silver bullet' that we expected it to be: many gaps and weaknesses as chinks in our armour remain to be exploited, and it still is possible for astute and experienced terrorists to use our society's strength and weaknesses against itself in a jujitsu-like, complex terrorism, just as Homer-Dixon (2009) predicted – an important issue I shall return to at the very end of this part. However, worse than the use of our 'silver bullets' against us by terrorists could be in stock for our open societies: our incessant 'wars' on terrorism, on crime, on drugs, on poverty etc. have given rise to a culture of fear and insecurity – a culture that is bound to transform not only our cities and the way they look and function, but our societies at large. I already touched upon this topic several times, for example in the introduction when I introduced the 'discourses of fear', and then again when I commented on the consequences (intended or not) of uncritically introducing the array of counter-terrorism technologies discussed in Part II. This unease regarding the consequences of relying on technology as a silver bullet in the fight against terrorism, and by extension also against crime in general, is rather common with the more critically inclined observers. And although some of them certainly are a bit too far on the alarmist side, many of them actually substantiate their queasiness with telling examples – which makes it rather difficult to simply dismiss them all as Luddites and naysayers. For this reason, this third and final part will offer a reflection of the consequences of an uncritical adoption of certain technologies as a silver bullet in the fight against terrorism for our societies.

At the core of this part lies what I labelled a 'critical' critical infrastructure protection approach in the introduction – an approach that does not get lost in mere technical descriptions, but keeps the following all-important question in mind: where does this all lead to? This 'this' includes not only our quest for ever more security as well as the consequences of adopting various technologies themselves, but also, and more importantly so, bigger-picture questions such as the impact of the 'Internet of Things' on our societies as far as that concerns us in our context: as I argued in the introduction, objects, or things, have agency as well since they tend to influence and change our behaviour. As such, their introduction may come – no, it will certainly come – with unintended consequences which are difficult to predict with any degree of precision. However, there are also bound to be 'intended' unintended consequences: consequences that 'we' as their targets are not necessarily aware of when a new technology is introduced, and which are not necessarily beneficial for 'us' as a whole – but possibly for some stake holders, be it public sector authorities or private sector enterprises offering solutions for all matters relating to safety and security – but I shall explain that below. Since this part admittedly depends on some crystal ball gazing, I shall juxtapose scholarly perspectives on current and foreseeable future developments regarding our societies with those envisioned in a sub-genre of science fiction known as cyberpunk, which had its heydays in the 1980s and early 1990s – as we shall see, there is a certain convergence between fact and fiction. But to begin with, and in order to give some necessary context to my following argumentation, some comments on our construction of 'citizenship' are unavoidable – after all, if I argue that counter-terrorism technologies, but low-tech and high-tech, have an impact on that, I need to define what citizenship actually means.

Chapter 11
The Quest for Silver Bullets: Implications for Our Construction of Citizenship

Abstract In this first chapter of the book's third part, I set the scene by discussing our conception and construction of citizenship. My core argument here is that new technologies, especially security-related technologies, change our view of what it means to be a citizen. In particular, I draw on the 'Agency of Things' as well as the 'Internet of Things' to highlight the relevance of technology in that regard. Since it is not only about technology per se but also the willingness to make use of it without too many critical questions asked, I also discuss some controversial views of politicians who seem to be willing to trade civil liberties for more security. I conclude this chapter by pointing out that we are already on a slippery slope in that regard, sleepwalking away from being citizens of liberal democracies towards being citizens of less liberal states driven by an 'us versus them' dichotomy much more pronounced than it is right now.

Keywords Agency of things · Internet of things · Citizenship · Exclusive citizenship · Inclusive citizenship · Nationalism · Universalism · Utilitarianism

When talking about the use of a variety of sophisticated technologies as silver bullets in the fight against terrorism and, by extension, violent crime as well, it is worthwhile to revisit one of Arthur C. Clarke's famous 'three laws' from his book *Profiles of the Future: An Inquiry into the Limits of the Possible.* His third law posits that "[any] sufficiently advanced technology is indistinguishable from magic" (Clarke 1973, 36). There is probably a lot to be said about crucial differences between technology, even the most sophisticated one, and magic, but the gist of his law seems to be quite right, as Cowper and Buerger (2003, 8) point out in their own work on police and augmented reality: "Many of today's new and emerging technologies appear magical to those ill equipped for rapid change." But there is an even more important argument to be made in this regard, which did not escape Cowper and Buerger's attention either: "These new technologies are changing not simply the world, but humanity's view of itself, giving people the ability to redefine the limitations and capabilities of the species, changing the nature of what it means to be human" (ibid.). Most certainly, at least in my opinion, these new technologies will also change our view of citizenship: what exactly does it mean to be a citizen?

P. Lehr, *Counter-Terrorism Technologies*, Advanced Sciences and Technologies for Security Applications, https://doi.org/10.1007/978-3-319-90924-0_11

Is it a right, or is it a duty? But before we discuss this crucial issue, we need to revisit two issues we already introduced and discussed in previous chapters: firstly, the 'Agency of Things'; and secondly, the vaunted 'Internet of Things' – both of which are of utmost relevance here as well.

11.1 The 'Agency of Things' and the 'Internet of Things': Reprise

Regarding the 'Agency of Things', the core argument is that not only people have agency, defined as the capacity of an actor to act in, or influence a, given environment – rather, objects, or 'things', can also wield such an influence. In the words of fellow anthropologist Bruno Latour as one of those scholars developing the concept further, "*any thing* [sic] that modifies a state of affairs by making a difference is an actor" (Latour 2005, 71 [emphasis in the original]). For the purpose of illustration, he offers some simple examples: "After all, there is hardly any doubt that kettles 'boil' water, knifes 'cut' meat, baskets 'hold' provisions, hammers 'hit' nails on the head [...]. Are those verbs not designating actions?" (ibid.). Of course, as he admits, the counter-argument could be that if one sticks to a rather narrow definition of 'actor' and 'agency', then one could argue that all these 'things' are nothing but passive tools gaining their 'agency' only via the intentional actions of humans. In a wider definition of agency however, according to which it depends on whether a 'thing' makes "a difference in the course of some other agent's action or not" (ibid.), then it is entirely possible to attribute agency to things as well. His most important argument on that matter, at least for our present context, is the following: "In addition to 'determining' and serving as a 'backdrop for human action', things might authorize, allow, afford, encourage, permit, suggest, influence, block, render possible, forbid, and so on" (Latour 2005, 72). This list should sound rather familiar by now since in the second part of this book, we examined a wide range of 'things' that can do exactly that: swipe cards equipped with RFID tags that can authorize or deny entry, smartphones equipped with facial or fingerprint recognition that allow or forbid access, scanners and sniffers that do the same with other technology and for a slightly different purpose, debit and credit cards that cover the 'afford or not' angle, smart apps that control and regulate our heating for us, remotely monitor the baby, or automatically order items if we run out of them, or make recommendations for restaurants, bars, shopping malls etc. on our smartphones or smart glasses – the latter soon equipped with Augmented Reality (AR) apps as well, in case the smart glasses gain mass acceptation and adoption this time (which was not the case for Google Glass in 2013–2015).

The smart apps bring us to the second issue of relevance here, and that is the Internet of Things. Simply put, the "Internet of Things' relates to the Internet-enabled high-tech gadgets now available that are meant to make our daily lives easier by improving overall accuracy and efficiency in conjunction with a reduced requirement for us directly interacting with them, made possible by the availability of cheap and miniaturized processing power. In a more scientific definition, and as McEwen and Cassimally put it, the Internet of Things is an equation of physical objects plus controllers, sensors and actuators plus the Internet (McEwen and Cassimally 2014, 11). Contrary to the passive low-tech tools Latour mentioned when he extended the concept of agency to 'any thing' as well, these high-tech tools in the shape of smart apps are programmed to be far more pro-active: depending on the type and operation system of your phone, we frequently notice that some 'helpful' suggestions, for example for bars, restaurants, shopping malls, theatres and the like, pop up without us asking for them, or even without us actually wanting them. What matters here is that there are at least two implications: first of all, the Internet of Things seems to have a massive impact on the Agency of Things by giving the high-tech ones of those 'things' more power to influence or even change our behaviour than the low-tech things 'enjoyed'; and secondly, the Internet of Things means that nowadays, and unlike the low-tech things, the high-tech ones are aware of our location, while oftentimes also having access to in-depth and detailed information about us as well. This implies that as convenient as all these new smart gadgets might be, they are not necessarily cost-free when it comes to civil liberties and human rights. As Williams and Selle (2016, 11) point out when they discuss the concept of 'Smart Cities': "There is, of course, the possibility that even smart cities might have their own dystopian underside where dense connectivity and transparency become intrusion, surveillance, and oppression." Williams and Selle's warning finally bring us back to the question already raised and then put on the back-burner: what exactly does it mean to be a citizen? And furthermore, since it should be clear by now that they have an impact on it, how do the 'Agency of Things' and the 'Internet of Things' as converging forces relate to our concept of citizenship?

11.2 Exclusionary vs. Inclusionary, Liberalism vs. Security: Some Comments on Citizenship

In order to discuss all these aspects, it is necessary to start with some definitions first. According to Merriam Webster's online dictionary, a 'citizen' can be defined simply as a) "an inhabitant of a city or a town; *especially:* one entitled to the rights and privileges of a freeman." Furthermore, it can be broken down to a) "a member of a state"; and b) "a native or naturalized person who owes allegiance to a government and is entitled to protection from it." 'Citizenship' as the second term then logically relates to the status of being a citizen. These definitions may well sound as trite as they are familiar, but they already hint at a certain tension: on the one hand, citizenship comes with rights and privileges, but with certain duties such as allegiance to a state on the other. Also, it tends to *include* certain sets of individuals, while *excluding* others. Regarding rights and privileges, we can go back all the way to Aristotle who, in the time of Greek city states, saw it as the right of the emancipated citizen (here defined as the citizen of such a city state, not of a territorial state) to freely participate in urban affairs, be they political or recreational. This was also the case from the Early Modern Age onwards, as expressed for example in the German motto "Stadtluft macht frei" or 'city air sets you free,' while in modern times, cities tend to be both melting pots and engines of innovations (Matzig 2018) – some examples to the contrary (such as US American rustbelt cities) notwithstanding.

The 'melting pot' idea however is not uncontroversial, as the much-discussed question of who should be included and who should not be included reveals. This question has resulted in many different theories and approaches, some more on the liberal side, some rather conservative (see for example Beiner 1995). To start with a liberal and inclusive one, I shall use the definition of democratic citizenship offered by Feliks Gross (1999, xi) who argued that it "extends human, political and civil rights to all inhabitants, regardless of race, religion, ethnicity, or culture. In a civic state, which is based on the concept of such citizenship, even foreigners are protected by the rule of law." However, as David Taylor alerts us,

> [The] political history of citizenship rights is inextricably linked to the history of Western European nationalism [with the effect that citizenship] becomes not only a process of a struggle for rights and entitlements in Western European nation states but also a struggle to reject claims of entitlement by those initially residing outside the core and, subsequently, of migrant and immigrant labour (Taylor 1994, 137).

In a nutshell, Taylor's argumentation signifies that even a Western liberal conception of citizenship can be seen as "a set of *inclusionary* and *exclusionary* practices, aimed at consolidating a particular set of social relations and of rights and entitlements" (ibid.). In a sense, and as I already mentioned several times, it comes with a built-in 'us versus them' dichotomy (on that, also see Anderson 2013, 29–47) that, amongst other themes, revolves around the preservation of security (Walters 2008, 167–168). Hence, thus far, the concept of citizenship already leaves us with two sets of tensions: *rights* versus *duties*, and *us* versus *them*. There also is a third set of tensions: that one between the interests of the individual citizen and the interests of the collective of all citizens, that is the nation or the state. From the former's perspective, it is mainly the rights that matter, that is the raft of civil liberties pertaining to the "Life, Liberty and the pursuit of Happiness", as the authors of the US Declaration of Independence of 4 July 1776 so eloquently and elegantly put it, to further explain that "to ensure these rights, Governments are instituted among Men, deriving their powers from the consent of the governed" (US Declaration of Independence). From the perspective of the state however (understood not as the government of the day but as an institution that transcends governments), the stress most certainly is on the duties – such as the duty to obey the law, to pay taxes, and to come to the defence of the state in times of war. Some states such as Belgium, Australia or Singapore even demand an active participation of the individual citizen in politics, at least as regards the duty to vote – a civic duty known as compulsory voting.

The issue of 'coming to the defence of the state' nowadays does not seem to only include joining the army, either voluntary or conscripted, in times of conventional war. Rather, it can be argued that in our current times of the Global War on Terror or GWoT (which has been officially renamed 'Global Contingency Operations' during the Obama administration, but the original 'Bushism' seems to stick), this has been extended to the duty of enabling the beleaguered state to pass ever more stringent laws geared to defeat this scourge, even if these laws impinge on our civil liberties and hamper our individual pursuit of happiness. This did not start with 9/11 or Al Qaeda, by the way: the threats posed by the Irish Republican Army (IRA) in the UK, the Red Army Faction (Rote Armee Fraktion, RAF) in Germany, the Red Brigades (Brigate Rosse, BR) in Italy, Direct Action (Action Directe, AD) in France or, at least to a certain extent, the Weather Underground in the USA, already started this securitization process in the 1970s (Fahrmeir 2007, 217). Of course, not all citizens saw a necessity to introduce specific anti-terror laws to combat what was, compared to Al Qaeda and ISIS, a rather minor threat. But the states in question usually managed to secure the acquiescence of the (silent) majority who continued to vote for the government of the day despite these measures. After all, it was usually not only an anonymous government or the proverbial Big Brother-like 'man' behind these anti-terrorism laws, but substantial parts of the political mainstream of the citizens as well, whose demands for more security against a faceless and sinister threat routinely were magnified by conservative tabloids and broadsheets – who also tended to over-hype and over-sensationalize the activities of the comparatively few terrorists – "six Red Army Faction members against sixty million West German citizens" in the case of Germany, as the German author and leading intellectual Heinrich Böll astutely summed it up (Aust 2008, 147–148). No less astutely, Malcolm Anderson (2008, 228) calls this knee-jerk reaction the 'politics of the latest outrage', as I already mentioned. Against the backdrop of the far more substantial threat posed by Al Qaeda and ISIS, this knee-jerk reaction is even more pronounced.

11.3 Kant or Mill: Universalism versus Utilitarianism

When I remarked that not all citizens tend to agree with the passing of ever-more stringent anti-terror laws ostentatiously meant to increase security but also prone to be misused and abused, I hinted at yet another tension – a tension not directly related to citizenship as such, but of eminent relevance for citizens: the tension between Universalism and Utilitarianism as guiding principles of all human action, including all political action. Universalism posits that there are universal ethics or moral truths that apply to everybody and must be maintained under all circumstances, regardless whether it is (politically) convenient or not. Since this philosophical approach goes back to Immanuel Kant, it is also known as Kantianism. On the other hand, Utilitarianism, with Jeremy Bentham and John Stuart Mill as its most famous proponents, posits that the greatest possible happiness of the greatest number of citizens should be the basis of all (political) action. Unlike the Kantian 'categorical imperative' that focusses on absolute moral truths, Universalism thus focusses on best possible outcomes – while tacitly acknowledging that catering for 'the greatest number' implies that there might be a minority whose happiness might diminish as a result of a certain political action. Unfortunately, it is in the nature of political action (politics defined as 'who get what, when and how' by Harold Lasswell, or as 'art of the possible' by Otto von Bismarck) that it is usually guided by compromises hammered out between various stakeholders, especially in democracies where many voices have to be heard, and many interests have to be accommodated. Hence, due to pragmatic considerations, Kantian absolute moral truths tend to fall by the wayside even in the best of circumstances.

In the context of 'greater happiness', Mill's interpretation of Utilitarianism is of special interest for us. He argues that "[it] is quite compatible with the principle of utility to recognize the fact, that some kinds of pleasure are more desirable and more valuable than others" (Mill 1987, 279). As he argues, there are 'lower pleasures' revolving for example around mere sensation and 'higher pleasures', such as safety and security – which brings us back to the modern state in a post-9/11 environment facing the threat posed by international terrorism and Al Qaeda or ISIS. Mill would probably concede that in such an environment, enhancing the safety and security of the greatest possible number of citizens via the passing of anti-terrorism laws and the introduction of new technologies to protect our open societies should have priority over the concerns of a minority, no matter how vocal, that these measures could impinge on civil liberties. Shabtai Shavit, former head of the Israeli intelligence service Mossad, expressed this utilitarian feeling rather pointedly, also quoting then UK Prime Minister Margaret Thatcher while doing so:

> Given a choice between being less democratic and surviving, and dying democratically, I prefer the first [option]. And in the intellectual discussion of the democratic dilemma we ultimately reach this point, and we must face up to it. Mrs. Thatcher said that in order to protect democracy, you sometimes have no choice but to use undemocratic means. (as quoted in Ganor 2005, 166–167)

Now, I assume that most of us tend to agree with Shavit's argument – after all, dying puts an end to our individual pursuit of happiness, while surviving a little less democratic probably does not. And if we happen to belong to the 'greater number' or to the category of 'us', we might not even feel the impact of Thatcher's (hopefully) few 'undemocratic means' introduced since they are not directed at 'us' but against 'them.' On balance, we might thus be inclined to shrug off certain limitations of our civil liberties with the usual 'if you don't have anything to hide you have nothing to fear' reflex. But in my point of view, and even without dragging Benjamin Franklin in the ring (I shall do so later), that brings us to a slippery slope, and back to the question asked before: how far can we go in order to protect us against terrorism – and more importantly, where will we end?

Interestingly, just on the day I was writing the paragraph above, the headlines of a short, two-columns newspaper article caught my eye: "Scottish criminals face no-go zones" it read on the front-page, and "Tagged criminals 'could be barred from entire cities'", it read on page 12 where the article itself appeared. It described how a 24/7 real-time tracking of (Scottish) criminals via electronic ankle tags could allow for setting up either 'exclusion zones', i.e. zones that said criminals would be barred from entering, or 'inclusion zones', that is zones the criminals would not be permitted from leaving (Green 2018). It also mentioned that the electronic monitoring of Scottish prisoners had been outsourced to private security contractor G4S (we already came across them above), which had just secured a two-year extension of their contract (ibid.). Entry denial, or exit denial in the case of inclusion zones, would still require a reaction by the police, alerted by an alarm triggered by monitoring stations reading the present location of electronic ankle tags via their RFID or GPS signal, depending on the device in question. In a sense, this could be seen as a very modern but extreme example of marshalling the powers of the both the 'Agency of Things' and the 'Internet of Things' in order to influence the behaviour of the individual in question. As infringements of civil liberties go, this one probably finds the support of most of us – after all, it only covers convicted criminals whose civil rights (especially so the right of free movement or free assembly) are restricted due to their custodial sentence in any case. But mind you that prisoners are not the only ones among us carrying certain wearables that continuously monitor their location: most of us voluntarily do the very same by wearing our access cards, smart phones or smart watches pretty much 24/7, as described above in Part II. This means no less that at last potentially, it would be possible for interested parties to marshal the capabilities of the Agency of Things and Internet of Things to monitor us and our movements 24/7 as well via GPS or RFID tracking, again depending on the device – and influencing our behaviour as well, either overtly or subtly.

Excluding us from accessing certain buildings for instance could be as easy as disabling our access card, and doors that used to magically open when we walked towards them would now stay closed. Denying us physical access to a facility would only be the start, however – we can easily spin this scenario on for our Oyster cards, debit and credit cards, or smart phones and all other smart gadgets that can be blocked. Most of the time, the exclusion from service would be a temporary one due to a computer failure or another mistake that could easily be explained and solved – the frequent travellers among us sure will have their own individual stories to tell about credit cards not working seemingly out of the blue when abroad and while trying to check in to a hotel: usually, that is due to an automatic security block because the location the card is used is not the normal one, which means one phone call usually is enough to re-establish service. But isn't there the gloomy 'what if' worst-case scenario on the back of one's mind if and when that happens, at least for some anxious minutes? Since we are law-abiding (I hope) citizens of Western liberal democracies, such a nightmarish 'what if' scenario seems to be as far removed from our existences as any science fiction movie is from reality. Nevertheless, by now we should have an inkling that (again, in theory) such 'what if' scenarios could be turned into reality right now and right here if there was an interest to do so – not necessarily malign, but just due to 'temporary' safety and security restrictions launched to protect the society as a whole against a clear and present danger posed by terrorists about to strike, or serious criminals about to do the same.

Now, my point is that this 'in theory' has already been turned into 'in practice' as well, due to our fear of terrorism. More precisely, I argue that counter-terrorism technology, both low-tech and high-tech variants, have already put us on a slippery slope that imperceptibly at first, but in the near future probably with an ever more rapid pace, leads us sleepwalking away from being citizens of liberal democracies towards being citizens of transfigured ones that thrive of an 'us versus them' dichotomy that will be more pronounced and profound than it is right now. There are several prongs of attack, so to speak – again without any malign intend behind it, and without us noticing. The first and most visible one is the gradual change of our vibrant cities and our vibrant open societies into less vibrant and open ones, in which citadels will become more frequent than they are now. The second and somewhat less visible one is the proliferation of surveillance technologies, which enable interested parties to keep tabs on our movements, if they are so inclined, in a cutting-edge version of Bentham's Panopticon. The third and probably least visible one is the emergence of private-sector enterprises which will (further) grow into quasi-state actor, reducing the 'real' state into an umpire state, functioning as an arbiter mediating conflicts, and thus being reduced to a role as 'primus inter pares', i.e. the first amongst equals. So, let us discuss these issues now in sequence.

Chapter 12
Archipelagos of Fear: CT Technology and the Securitisation of Everyday Life

Abstract In this chapter, I re-examine the transformation of our cities under the impression of recent terrorist attacks from a critical perspective. I argue that a 'discourse of fear' enables a process that turns ever more of our public spaces into 'safe spaces' which are essentially 'quasi-public' only – quasi-public in the sense that they can be accessed only by those citizens fortunate enough to have the right credentials, thus excluding or 'othering' all those we deem to be 'undesirables', however defined. I point out that this exclusion already is a common practice – and not necessarily connected to the threat of terrorism. Rather, in my view a 'hostile architecture' has emerged that targets everyone who does not fit in. To defend my point of view, I discuss concepts such as 'defensible space', 'architecture of fear' and 'archipelagos of fear' in the shape of loosely connected inner-cities citadels and gated communities in the suburbs.

Keywords Barricades · Citadels · Gated communities · Protected zones · Architecture of fear · Hostile architecture · Defensible space · Archipelagos of fear · Bubble laws

Regarding the changing nature of our cities in general and of our vibrant open societies in particular, let us start with a little thought exercise: think of our Western cities, or the one you are most familiar with, and take note of the mental images that pop up in your mind. I take it that not all of them are positive: for example, if you are a Londoner, you might have thought of a packed tube train at rush hour, while if you were from Los Angeles, the seemingly endless commute along the highways from the suburbs to the centre (or centres in this particular case) that can take hours. There is the problem of congestion at least twice per day when commuters rush in and out of the city, there is smog and air pollution as well, and there are the usual armies of panhandlers, buskers, and so-called chuggers (charity muggers) forming an obstacle course that needs to be negotiated. Fair point. However, after an initial grumpiness, a more positive image of the city might appear as well: the hustle and bustle, the lights, the numerous attractions in the shape of parks, cafes, museums, theatres, cinemas, sports stadiums, or the proverbial city life in general that was

P. Lehr, *Counter-Terrorism Technologies*, Advanced Sciences and Technologies for Security Applications, https://doi.org/10.1007/978-3-319-90924-0_12

celebrated, amongst others, in Petula Clark's 1965 hit *Downtown*. Anthropologist Teresa Caldeira captures the city life in its positive aspects quite nicely:

> [In] Western cities, the modern experience of urban life is widely held to include the [...] presence of people from different social backgrounds strolling and gazing at others, looking at store windows, shopping, sitting in cafes, joining political demonstrations, appropriating the streets for their festivals and celebrations, and using spaces especially designed for the entertainment of the masses [...] These are elements associated with modern life in capitalist cities at least since the remodelling of Paris by Baron Haussmann in the second half of the nineteenth century. [...] The flaneur described by Baudelaire and the consumer of the new department stores became symbols of the modern use of urban public space, as Paris became a prototype of the modern city. (Caldeira 2000, 299)

In a (historical) sense, she is overplaying this novelty – after all, city life in one respect or the other emerged together with the cities themselves. The main streets may not necessarily have been called 'boulevards' then, store windows still were notably absent, and cafes did not yet exist either. There were however bazaars, markets and fairs, as there were taverns, guest houses or brothels, for that matter. In any halfway decent ancient Greek and Roman city, there was at least one major forum, to serve both as a market place but also as a public meeting venue for discussions. There were public baths, and there were sports venues – some of which also provided convenient venues to address the masses with political speeches or agitation, as happened in the hippodrome of Constantinople in the year 532 when the normally hostile factions supporting various chariot racing teams suddenly buried their hatches to turn against Emperor Justinian instead – an incident known as the Nika riots ('Nika' means 'Victory'). Be that as it may, even if we can criticise Caldeira's for her undue focus on 'modern' urban city life, her argumentation as such still is sound and convincing: it implies that in essence, urban life depends on the existence of public spaces where all the interactions just described can take place.

12.1 Barricades and Citadels Revisited: The Changing Nature of Public Spaces

City life obviously depends on the availability of public spaces where it can take place. Unfortunately, the public spaces just described now are under threat – not necessarily, or at least not solely, from terrorists trying to mow us down with trucks as in Nice or Berlin, or kill us with assault rifles and hand grenades as in Mumbai and Paris. In Chap. 9, I already drew attention to the shrinking of our public spaces, and the change which many of them undergo – a change that results in them being turned either into only 'quasi-public' spaces (i.e., spaces that now can only be accessed by presenting the right credentials, for example an ID) or into 'closed' spaces (i.e. former public spaces that now are off limits for the public). I illustrated these arguments with some examples from various Western cities, also highlighting the role of private sector actors within these developments – a role that also includes securing these converted spaces with their own private (quasi) police forces. Kaplan (2000, 83–84) already commented on that process nearly twenty years ago:

> The number of residential communities with defended perimeters that have been built by corporations went from one thousand in the early 1960s to more than eighty thousand by the mid-1980s, with continued dramatic increases in the 1990s. [...] Then there are malls, with their own rules and security forces, as opposed to public streets; private health clubs as opposed to public playgrounds; incorporated suburbs with strict zoning; and other mundane aspects of daily existence in which – perhaps without realizing it, because the changes have been so gradual – we opt out of the public sphere and the 'social contract' for the sake of a protected setting.

With regard to the undemocratic character of this gradual and oftentimes imperceptible change, Norris and Armstrong (1999, 8) offer a very similar argument, pointing out that public spaces essentially are "democratic spaces in the sense that access to them is not dependent on status, but guaranteed by virtue of being a citizen." These 'democratic spaces' do not only fulfil the function of 'recreational spaces', but also, according to Flusty (1994, 12) as "places for the exchange of goods and ideas including those too novel, threatening or unpopular [and] they are visible staging areas for alternative and oppositional as well as establishmentarian actions." Barricading them however tends to strip them of this character, and to turn them into both pseudo-democratic and pseudo-public spaces. Norris and Armstrong (1999, 8) describe this gradual change quite forcefully:

> Increasingly [...] this public space is being reconstituted, not as an arena for democratic interaction, but as the sight of mass consumption. Individuals are recast as consumers rather than citizens, as potential harbingers of profit, rather than bearers of rights.

This consumerization of citizens with upmarket shopping malls pretty much assuming the role of modern temples however is only the tip of the iceberg of changes. On the basis of the broader understanding of technology (following the Oxford Online Dictionary) as the "application of scientific knowledge for practical purposes" that was mentioned in the introduction to Part 2, Kaplan's comment quoted above also astutely illuminates the wider implications of the adoption of modern technology in the quest for enhanced security: this seemingly innocuous 'win-win' strategy leads to fundamental changes within our society without us noticing. In my opinion, one of the main reasons behind this is due to the fact that this 'win-win' strategy is based on a solely technocratic understanding of security which forms the core of the uncritical 'critical infrastructure protection' that I already flagged up. To return to Kaplan, he concludes his argumentation by noting that US cities seem to be "re-emerging as Singapores, with corporate enclaves that are dedicated to global business and defended by private security firms adjacent to heavily zoned suburbs" (Kaplan 2000, 84). Flusty (1994, 37) drastically illustrates the wider implications of this process of 'citadelization' on the example of Los Angeles in an eloquent argument that deserves to be quoted at length:

> Los Angeles is undergoing the invention and installation, component by component, of physical infrastructure engendering electronically linked islands of privilege embedded in a police state. If left unchecked [...] Los Angeles will become a city consisting of numerous fortified cores of private space, each augmented by more permeable outer perimeters of

> contorted paths, lights, motion detectors and video cameras projecting into the public realm of sidewalks and streets. The public streets will become little more than interstitial space to these fortified private cores. They will themselves be fragmented by selective barricading and monitored by [surveillance] cameras overlapping each private space's permeable outer perimeter.

We could of course question with Norris and Armstrong (1999, 9) in how far the particular experiences of Los Angeles should be seen as applicable to other US cities, or to European cities at that: as Davis (2006b, 31) for example points out, "[in] the First World [...] there is an archetypal distinction between 'donut'-shaped American cities, with poor people concentrated in derelict cores and inner suburbs, and European 'saucer' cities, with immigrant and unemployed marooned in high-rise housing on the urban outskirts." Hence, and even though the problems zones of the French banlieues have already been discussed, and the emergence of 'no go' zones for the police in certain cities been commented on as well, different patterns of urban defensive geographies will certainly arise. It is also true that Flusty's view of the future is only one possible way to look at it, and a rather pessimistic one at that. Nevertheless, and despite these two caveats, the lessons learned in Los Angeles still are applicable to, and relevant for, different Western urban areas – especially so since many cities, including 'old' European ones with a long and proud history, seem to be shrinking and to undergo a process of 'de-urbanization' for a variety of reasons, such as an ageing (and thus decreasing) population, the disappearance of those core industries (now no longer in demand) that led to the growth of certain cities in the first place, or the depletion of the natural resources that turned bustling boom towns into ghost towns (Biswas et al. 2018).[1]

Shrinking populations apart (which come with their own problems which are not relevant in our context), a logical side effect of this de-urbanisation can be seen in the rapid growth in population of those small towns and villages in easy travel distance to those cities fortunate enough to host the industries of the day (such as finance, IT, biochemistry), but whose booming cores are no longer affordable even to the upper-middle classes due to the sky-high real estate prices or exorbitant rents demanded – London or Munich would be examples for that. The equally logical knock-on effect of this process then is the gradual process of 'pricing out' the locals since the real estate prices and the rents are rising there too due to the influx of relatively well-to-do domestic migrants. Again, the regions around London and Munich would be examples for this development. It could be argued that this process might gradually weaken the differences between US American 'donut cities' and Western European 'saucer cities', but there is no need to resolve this issue here since it does not really matter – we can leave this to specialists in urban geography. What does matter however is that due to this de-urbanisation, a sub-urban fringe of Flusty's 'fortified cores of private space' populated by the (relatively) affluent parts of the population might find themselves cheek-by-jowl with urban problem zones featuring

[1] An interesting article on water networks in Eastern Germany sheds some light on the knock-on effects of that, for example by quoting a planner from municipal water company of Leipzig as saying that "[we] have the demand of 1945, with a network of 1990" (Naumann and Bernt 2009, 464).

a high rate of crime, and thus as (fortified) islands of prosperity in a sea of destitution. In a worst-case scenario, this could create quite a dystopian environment, as envisioned for example in a sub-genre of science fiction known as Cyber Punk.

12.2 Security as a 'Culture of the Absurd': Bubble Laws, Protected Zones, and Totally Protected Zones

At the time of writing, dystopias as described in the Cyber Punk novels, or in movies such as *Blade Runner,* do not yet exist, and it is questionable whether they will ever emerge: as mentioned in the introduction, our ability to predict the future with any degree of precision is quite limited. What we can do in order to get at least an impression of what could possibly happen to our Western cities in the foreseeable future is to take a look outside our own bubble: to the various megalopolises and urban sprawls of Africa, Asia and Latin America. Doing so actually alerts us to the fact that 'off world colonies' as predicted by Philip K Dick in his famous novel *Do Androids Dream of Electric Sheep?* (on which Ridley Scott's movie *Blade Runner* is based) are not necessarily as far-fetched as one might think – and already very near to the Cyber Punk idea of 'burbclaves' (a portmanteau coined from 'suburban' and 'enclaves') and (semi-) sovereign enclaves run by private companies and protected by private security companies or mercenary armies as envisioned in Neal Stephenson's Cyber Punk novel *Snow Crash.*

From the perspective of urban geography, many studies of suburban fringes (or inner-city parts) of rapidly growing megalopolises have already been carried out, and there is no dearth of articles ranging from the rather general to the very specialized. Fortunately, several easily accessible monographs and edited volumes also exist that bring much of this knowledge together – one of the being Mike Davis' *Planet of Slums* (Davis 2006b). In his chapter 'Haussmann in the Tropics', Davis discusses several cities and their suburbs and enclaves for the middle classes and the rich – for example Bangalore in India, the Philippines' capital (Metro) Manila, and South Africa's capital Johannesburg, also referring to Philip K. Dick's 'off worlds' while doing so. Of particular interest here is his focus on matters of security, bordering on the paranoia, within these gated communities patterned after Californian towns such as Palo Alto and Sunnyvale due to their lifestyles (Davis 2006b, 116). Since 2000, I visited similar gated communities in India's sprawling capital New Delhi as well as in the equally sprawling port city of Mumbai (Bombay) – albeit solely to live there during my research trips, not to research them. Nevertheless, I could not escape noticing that over the years, many of them seem to have changed: while in the beginning, gate communities tended to just cluster around openly accessible market centres and/or shopping strips, the modern and much more upmarket variants tend to include such amenities and more within their walls – including swimming pools, gyms, cinemas and nightclubs, possibly even a hospital (for Indonesia, see Leisch 2002, 345). As such, these evolved gated communities now truly function as autarkic sub-sets of the city, and as 'bubbles' or 'off world communities' in their own right. Even more importantly, many of these gated communities, especially those for the middle classes who may well be 'upwardly mobile' as they would claim but still cannot yet leave the city for good and for the more secure suburbs, also ramped up their security, and quite drastically so: while in my times, largely amateur security guards and night watchmen were mainly equipped with long bamboo sticks (also known as 'lathis') and the odd obsolete Enfield rifles mentioned in the first part, their modern variants are far more professional in their demeanour, their training, and their equipment, well on par with comparable Western technologies.

In a sense, many of these gated communities, at least those populated by the middle classes, seemed to have morphed from 'protected zones to 'totally protected zones', as Leisch argues in the case of two suburbs of Jakarta, Lippo Karawaci and Bumi Serpong Damai (Leisch 2002, 344). On the basis of a series of interviews he conducted in 1999, Leisch found that "security was the main or one of the main reasons to move to the new towns" (Leisch 2002, 348). In the case of upper class residential areas such as Menteng in Central Jakarta, security is even higher on the agenda: here, every dwelling can be seen as an individual 'totally protected zone' or as a veritable citadel in its own right guarded by walls that tend to grow in height after every successful break-in – a luxury that the middle class cannot afford, hence the need to cluster in a gated community (Leisch 2002, 343–344). In any case, Davis (2006b, 116–117) argues that such fortified, totally protected communities are a common feature of the Third World's urban area – features that can nowadays also be found in many parts of the First World, however. Amongst others, Davis draws attention to the work of Erhard Berner to illustrate this argumentation. Berner's main research interests was to assess how wealthy suburbs situated in a sea of poverty protect and defend themselves – which is why he chose the Loyola Heights as one of Metro Manila's most exclusive district. Berner describes this district's security features as follows:

> An elaborate system of iron gates, roadblocks and checkpoints demarcates the boundaries of the area and cuts it off from the rest of the city, at least at nighttime. The threats to life, limb, and property are the overwhelming common concern of the wealthy residents. Houses are turned into virtual fortresses by surrounding them with high walls topped by glass shards, barbed wire, and heavy iron bars on all windows. (Berner 1998, 163; also quoted in Davis 2006b, 116)

More than two decades after Berner wrote this description, we could probably ask in how far Loyola Heights differ from many of the no less protected and defended wealthy gated communities in the suburbs of Western cities – but the sheer fact that we can actually raise this question indicates just how far we already moved towards dystopia. Be that as it may (we shall return to this question again later on), Davis also refers to the works of anthropologist André Czeglédy who travelled through South Africa's capital Johannesburg, visiting its northern suburbs populated mainly by affluent white citizens with a research agenda very similar to Berner. Indeed, Czeglédy finds suburbs akin to the 'totally protected zones' just mentioned, and a "security [that] has become a culture of the absurd" (Davis 2006b, 117):

> The high perimeter walls are often topped by metal spikes, razor wire, and more recently, electrified wiring connected to emergency alarms. In conjunction with portable 'panic button' devices, the house alarms are electronically connected to 'armed response' security companies. The surreal nature of such implicit violence was highlighted in my mind one day when walking with a colleague in Westdene, one of the more middle-class neigborhoods of the Northern suburbs. On the streets was parked a minivan from a local security company that boasted in large letters on the vehicle's side panel that they respond with 'firearms and explosives.' Explosives? (Czeglédy 2003, 36; also quoted in Davis 2006b, 117)

Such modern suburban dystopias indeed betray distinct features of an 'architecture of fear', as Tunde Agbola termed it in the case of well-protected, fortified wealthy neighbourhoods in the Nigerian capital Lagos (Agbola 1997). As he points out, "whatever the spatial or temporal distribution of crime, the immediate response of people to this malaise is fear" (Agbola 1997, 1.3), with the effect that the "general appearance of most residential buildings in the low to medium density areas in Lagos is like that of fortresses/strongholds" (Agbola 1997, 3.16). The most frequent security measures adopted were, according to his study, high concrete fences, steel barriers, illumination of properties, burglary proofing of windows and the use of security guards, especially in areas of a low to medium density, while high-tech security measures such as CCTV usually were found only in low density, high income residential areas (Agbola 1997, 3.17–23). Of particular significance in our context on the increasing citadelization of cities is his finding that "[some] residential buildings were not initially developed or built with the security found in them [but] were remodelled in response to the increasing menace of different types of urban violence" (Agbola 1997, 3.24). In my opinion, a similar trend can also be seen in those parts of Western cities which were so far not as much affected by serious crime as Lagos, Johannesburg, Metro Manila or São Paulo, but are now reeling under the seemingly unstoppable onslaught of terrorism.

12.3 'There Be Lions': From 'Architectures of Fear' to 'Archipelagos of Fear'

If seen and evaluated separately and on a case by case basis, gated communities and other manifestations of citadels can well be called 'architectures of fear', as we have just seen. However, when seen in conjunction, and as a whole, these citadels show features of an 'archipelago of fear' as I would like to call it: well-protected islands of (relative) affluence in a sea of deprivation and poverty, or "[fortresses] of office blocks whose streets empty at dusk" (Kaplan 2000, 85) in the case of inner cities. Seeing them this way raises some interesting questions. The obvious first question is how to connect them without leaving the secured environment – after all, one still has to venture out, which entails the risk (as overblown as it may be) to chance upon terrorists eager to mow us down or to gun us down, on top of having to run the gauntlet of all those people unfortunate enough not to afford to live in a gated community, and also potentially and rather vaguely posing a danger to 'us' – at least those of us who cannot afford to hire a helicopter for the daily commute.

Somewhat ironically, commuting between these corporate enclaves, zoned suburbs and privately-owned and controlled/policed malls is reminiscent to a potentially dangerous voyage not necessarily through uncharted territories that used to be denoted by cartographers of old either with 'there be lions' (hic sunt leones) or, even more dramatically, 'there be dragons' (hic sunt dragones), but most definitely through unpoliced and, thus, unsafe, territories. An advert for a SUV illustrates that very well: as the sales pitch goes, "[the] harshest roads are not necessarily located out in the boonies." And to drive it home to the potential buyer what is meant by that, the SUV is depicted in a dark, threatening, obviously dilapidated and run-down urban environment – an inner-city slum, so to speak, or at least a street of a particularly 'mean city' (in Graham (2010, 308). Graham hence suggests that "SUVs can be understood as mobile 'capsular' technologies, designed to offer autonomous neoliberal subjects the fantasy of complete individual control and total libertarian secession from the social and public spaces of city life – spaces which, because they exist beyond the cocoon of the interior, become residualized" (Graham 2010, 309). Gunster (2004, 23; also quoted by Graham 2010, 309), offers a similar argument, stating that the "[incessant] celebrations of a luxurious interior defended by an armoured shell champion the mobile and aggressive privatization of public space in which those with wealth and resources can use and enjoy the commons while maintaining complete control over their own personal environment." This 'cocoon of the interior' that Graham mentions is nothing else but an assemblage of the various types of citadels that were discussed: gated communities and condominiums, corporate enclaves, malls, plazas, private-owned or state-owned transportation hubs in the shape mostly of airports, lesser so of rail or metro stations due to the fact that most of them still are outside of this cocoon.

Interestingly, this particular 'citadel on wheels' even has a potentially offensive character, as Gunster points out, invoking Adorno who quipped in his *Minima Moralia:* "And which driver is not tempted, merely by the power of his engine, to wipe out the vermin of the street, pedestrians, children and cyclists?" (Gunster 2004, 4). Gunster takes great care to drive home this not-so-hidden belligerent trait of SUVs, arguing that "[as] the fear of crime, however irrational, has risen in lockstep with the intensification of violence in the mass media, the SUV offers itself as an ideal technology for armouring the self against the perceived dangers that lurk outside" (Gunster 2004, 15). He draws attention on French anthropologist Clotaire Rapaille, who brings the argument to a rather pointed and controversial conclusion that however goes well with Adorno's remark:

> I think we're going back to medieval times, and you can see that in that we live in ghettos with gates and private armies. SUVs are exactly that, they are armored cars for the battlefield. (as quoted in Bradsher 2002, 97)

Quite obviously given their price tag, focusing on cars such as SUVs indicates that we are largely talking about middle-class citizens. Citizens above them in the social pecking order, i.e. members of the upper-class, might find this commute even less threatening since their personal wealth allows them to firmly remain within the cocoon at all times, for example using helicopters for the essential trips through exposed parts of the city, while also more often than not having access to highly secured private sections of airports, including private jets – for example those of the corporations they work for. On the other hand, all those citizens who for one reason or another do not have an SUV at their disposal might deem their personal daily commute through exposed parts of the city much more adventurous: in order to get from A to B, they will usually have to take a mass rapid transport system such as a train, a metro, or a bus – thus required to visit bus terminal, underground and train stations which are harder to secure than airports due to the high volume of passengers, as discussed in the previous part. I should however hasten to add (as Graham does) that this 'cocoon of interior' in the more precise sense of 'cocoon of safety and security' is more of an imaginary one than real. After all, a constantly recalibrating GPS system could easily lead this mobile cocoon astray and expose its passengers to nasty incidents such as car-jacks or kidnappings.

On a less-dramatic level, and as soon as they get out of their respective mobile cocoons, most citizens will still have to 'mingle' with those unfortunate enough to be less well-off – such as beggars, homeless persons, buskers, street artists, pimps, touts and other marginalized individuals trying to eke out a living on the streets. Furthermore, the streets and other parts of our public spaces also are the venue of choice for vociferous protest movements whose members can be, at times, very pushy and aggressive. Hence, American law attempted to create additional cocoons of security or 'floating buffer zones/floating bubbles' by passing so-called 'bubble laws': laws that make it illegal to knowingly approach an individual's personal zone, with eight feet seen as "the proper distance" (Mitchell 2005, 83). Mitchell comments on the implications of this trend within US American law as follows:

> The importance of the issues at stake should not be minimized. City spaces are those places where the public comes together in its diversity, and where, presumably, the interaction of difference helps create the possibility for democratic transformation. [...] But the creation of buffer zones and floating bubbles seems to be mitigating exactly this ideal. Or perhaps more accurately, it is promoting a form of access to public spaces and forums that radically transforms possibilities for association in them. (Mitchell 2005, 86)

Furthermore, Mitchell astutely connects this trend and its implications to the role surveillance technologies play in this regard:

> Closed circuit television and other sophisticated surveillance technologies [...], coupled with a general rise of [...] middle and upper class 'revanchism' that has targeted homeless and other marginalized people for 'zero tolerance' policing [...], reinforces this trend towards promoting access (for some) to public space by closely regulating the form that association may take. (Mitchell 2005, 87)

Mitchell's argumentation reminds us that rebuilding our cities and rearranging (or closing) our public spaces is only the visible tip of the iceberg here: the 'silicon-based' and potentially very intrusive surveillance measures such as biometrics, 'smart' CCTV or real-time data mining and data acquisition are bound to change our societies even more – I shall come back to this topic in the following chapter.

The second question that comes to mind in the context of these 'archipelagos of fear' is about the implications for citizenship – a topic that we already touched upon several times. After all, it is unlikely that 'all of us' will be able to move to gated communities or other manifestations of citadels, or that 'all of us' will be treated equally in the eyes of those responsible for surveillance, observation and identification tasks. To start with the inconvenient fact that not all of us can move into a gated community or a citadel, this also has consequences for us and our conception of citizenship in the shape of an increase in segregation and quartering, as Peter Marcuse points out:

> [The] polarization by income, by occupation, and by race that is an ongoing process in central cities will accentuate a further developing pattern: a barricading of segregated spaces. [...] Those able to move out of town or to barricaded citadels will do so; those unable to do so will remain behind. The difference between the two groups will be both income and race related, with sharper dividing lines between and among groups. (Marcuse 2004, 271–272)

This has obvious implications for the question discussed above whether interlinked citadels will emerge or not, understood as a neo-urban layer of safe and secure zones that has only little contact with the unsecured parts of the urban environment – parts which then can easily be left behind because they do not matter any longer. Regarding the issue of (un-) equality, there already are parts in our societies which are manifestly not treated equally even though they are proper citizens entitled to the same rights as everybody else. I am alluding here to the well-known practice of 'racial profiling' that, depending on the terrorist threat level of the day, makes certain individuals more likely to be stopped and searched than others. Against the backdrop of Islamist terrorism for example, young Muslim males of Asian ethnic descent are far more often stopped and searched than, say, middle-aged white males or white octogenarians. As one police officer once bluntly told me about this much maligned and officially non-existing policy, 'Mancunian lorry drivers or Liverpudlian grandmothers are unlikely to blow themselves up for ISIS.' That ISIS is aware of that and thus tries to recruit individuals who do not look like the stereotypical Islamist suicide bomber in order to outmanoeuvre this thinking, just like the PFLP-EO used three operatives from the Japanese Red Army for the Lod Airport attack of 1972 to do the same with the Israeli racial profiling, is a moot point here. Of more importance in this context is that the biometrics discussed working in conjunction with access controls could automatize this profiling by simply denying certain groups within a society access to certain areas – or at least use biometric IDs in conjunction with smart CCTV systems to keep an eye on their movements. Steps in this direction have already been taken – ostentatiously for benign reasons.

A (non-Western) example would be the biometric ID cards rolled out for slum dwellers in Chennai, India during 2012 and in Mumbai and Thane, also in India, in December 2015. These cards that have a Unique Identification Number (UID) and include fingerprint patterns are meant to be proof of residence that entitles the carrier to a range of benefits including food subsidies, and to 'cut out the middleman', that is to avoid corruption. Critics however were quick to point out that there was little debate only about the introduction of such cards, and that the information collected could be misused – for example in the federal state of Jammu and Kashmir that is ripe with insurgency and acts of terrorism, or in Gujarat that has a history of communalist clashes between Hindus and Muslims. However, so far it seems that the cards do not include information about one's religion or caste – crucial information that would be necessary for a profound 'othering' of 'undesirables' (Patnaik 2011). But again, this example shows how easy it would be to tweak well-meant welfare-facilitating measures into less well-meant surveillance measures either for inclusion or exclusion purposes, just like those discussed in the case of Scottish prisoners wearing electronic ankle tags, as discussed above.

As concerns the profound 'othering' of 'undesirables', it should be noted that even in Western liberal democracies, this already is a common practice – not only on the front of racial profiling, and also unconnected to the threat of terrorism. Rather, a 'hostile architecture' has developed that aims at basically everyone who does not fit in, and who thus have to 'convinced' to move on in one way or another. Examples of such hostile architecture, also known as 'unpleasant design', 'hostile design' 'defensive urban design' or 'defensive architecture' (see for example Savicic and Savic 2016; Minton 2017), would be metal spikes in front of shops, or iron bars across benches, in order to prevent the homeless to sleep there. The incessant piping of music in facilities that have to be open 24/7, such as railway stations, is another example – and so are the devices that emit a mosquito-like sound inaudible to all but teenagers, just to keep them from using the thus protected space. Light has been used for similar purposes – for example in the shape of flood lights drawing a highly visible wall of light between public and (semi-) private space, in the shape of "pink lighting that apparently draws attention to acne […] to prevent teenagers from congregating" (Beanland 2018), or in the shape of blue lighting in spaces frequented by drugs users – blue light allegedly makes it harder to inject drugs due to the fact that veins are rendered less visible (Crabtree et al. 2013). Interestingly, as Dean Harvey points out, it is not necessarily the state or the city council which is the driving force behind this 'unpleasant design': "The public like to blame councils and developers but the problem is more ingrained with attitudes to children, youth and people on the edge of society" (as quoted in Beanland 2018).

The discussion so far reconfirms my argument that the search for more security is neither tied to nor triggered by the rather novel threat posed by terrorism – again, I agree with Mueller (2006) that this threat is overblown – but is a much wider phenomenon. I argue however that this novel threat, especially the one posed by Al Qaeda and ISIS, further accelerated the already existing trend towards hardening our cities, also offering yet another compelling reason why this has to be done, and why counter-terrorism technology has to come to the rescue. Anderson's 'politics of the latest outrage' most certainly play a prominent role in that regard. In a less dramatic sense, this hostile architecture as part and parcel of the discourse of fear simply is a continuation of Oscar Newman's already mentioned 'defensible space': after all, the intention behind that was offering an alternative to a "total lockup and self-restriction of movement [or] a self-imposed curfew and police state" by way of "restructuring the residential environment of our cities so that they can again become liveable and controlled, controlled not by police but by a community of people sharing a common terrain" (Newman 1972, 2). What is liveable or not obviously depends on the eyes of the beholder – which is not a trivial point, but one that raises the question of who are our cities for? After having discussed zero tolerance policing, burbclaves, architectures of fear, archipelagos of fear and hostile architecture, this question brings us back to where we started: the implications of modern security technology on our concept of citizenship, and our understanding of democracy.

Chapter 13
Undemocratic Means: The Rise of the Surveillance State

Abstract In Chap. 11, I already raised the 'what if' question regarding the imagined slippery slope from liberal democracies to more authoritarian and far less liberal or democratic states. In this chapter, I show where this 'what if' journey towards ever more safety and security could eventually lead by discussing the Chinese government's current plans to roll out a 'Social Credit Score' system that assigns each Chinese citizen an individual 'social score' – a score that can go up and down depending on their action and how 'trustworthy' they are. I argue that this total surveillance goes for beyond 'nudging' as employed in Western democracies, but that for those who score high, it actually does not look like surveillance but like a game – hence, we basically witness the 'gamification' of social control. I then compare these developments to surveillance measures already existing in the West and conclude that we are actually not that far removed from China's version of Huxley's 'Brave New World.'

Keywords Big data · Smart CCTV · Megvii Inc. · Skynet · Dragonfly eyes · Nudging · Gamification of social control · Social credit system · Total information awareness · Data-mining · Identification bias

I already raised the 'what if' question regarding the imagined slippery slope from liberal democracies to more authoritarian and far less liberal or democratic states, referring to former British Prime Minister Margaret Thatcher's quip that in order to protect democracy, sometimes you have no choice but to use undemocratic means. From a civil liberties and human rights perspective, this is the proverbial elephant in the room since these undemocratic means ostentatiously adopted to save us from the threat of terrorism have the potential to change our societies even more profoundly than the architectures of fear just discussed. After all, it can be argued that such measures make good use of the discourse of fear and the 'us versus them' dichotomies that it deepens to change the nature of our democracies as such, for example by turning liberal democracies into illiberal ones. At the moment, this still can largely be seen as a 'what if' scenario that may not necessarily come true – although, as we shall see below, it is not a purely speculative one any longer either. Where this 'what if' journey towards ever more safety and security could eventually lead to can be

P. Lehr, *Counter-Terrorism Technologies*, Advanced Sciences and Technologies for Security Applications, https://doi.org/10.1007/978-3-319-90924-0_13

seen in an admittedly non-democratic state some steps removed from the Western liberal democracies discussed here, and that is the PR China. So, let us start there, to then look at possible implications for us, our own open societies, and for our liberal democracies as such.

13.1 'Dragonfly Eyes' or 'Keeping Trust is Glorious': The 'Brave New World', Chinese Version

As German journalist Kai Strittmatter (2018) sees it, the Chinese government, not encumbered by civil liberties issues and pressure groups fighting for them, has discovered Big Data and Artificial Intelligence as magical weapons when it comes to establish a country-wide complete surveillance of its citizens and their activities. To start with, it can bank on the help of Western high-tech companies eager to get access to the Chinese market as well, even if this means to accept conditions and limitations regarding their operations. For instance, Apple drew quite a lot of criticism when it decided to host "iCloud data inside China despite concerns from human rights activists that authorities could use information to track down dissidents" (Barnes 2018). As Tom Barnes points out, this is "the first time the company has stored cryptographic keys needed to unlock iCloud accounts outside of the United States" (ibid.). Apple's defence that it has "been forced to comply with recently introduced Chinese laws that require cloud services for Chinese citizens must be operated by Chinese firms" (ibid.) may well be a valid one if seen from the legal and commercial side of things, but it does not address the concerns regarding potential human rights violations, such as the merciless prosecution of dissidents. Apple is not alone when it comes to complying with Chinese demands regarding the handing over of sensitive data – Yahoo made the same experience years ago (ibid.).

But this virtual snooping on its citizens with the help of technological giants such as Apple is not all there is: according to government plans, as many as 600 million CCTV cameras are to be deployed until 2020 in order to be able to actively snoop on citizens and their activities as well – Smart CCTV systems, to be precise: they are meant to be capable of reading and identifying the faces of passers-by via sophisticated facial recognition software, plus adding available information (name, address, ID or passport number) on those flagged up by the system in order to let an officer in the control room decide whether or not to take action, for example by guiding a police patrol towards a suspect. Encouraged by the government and its ambitious plans, a rising number of high-tech firms currently work on ever smarter CCTV systems that are meant to go beyond mere facial recognition to do all that we discussed in the previous part and much more. One of them is Megvii Inc. (also known as Face ++), a start-up founded in 2011 which boasts that it is doing no less than developing the 'eyes of the city', and intelligent ones at that, able to identify and, if need be, follow individuals from camera to camera until they have reached their destination. For that purpose, Megvii can draw on the Chinese Ministry of

Public Security's database, according to one source containing the files of about 1.3 billion Chinese (L. Y. Chen 2017), and to another, a more credible 700 millions of Chinese aged 16 and more since they are usually issued with a photo ID when reaching this age (The Economist 2017).

On its webpage, the firm states that its cutting-edge software already is capable of supporting networks of 50,000 to 100,000 CCTV systems – systems allegedly so sophisticated that a spokesperson of the firm, Xie Yinan, wonders whether crime actually still pays given the near certainty that criminals will be detected, observed and identified while still in the process of carrying out their crime (as quoted in Strittmatter 2018, translation mine). This may well be a tad too optimistic at the moment, but fact is that during the 2017 'Octoberfest' of Qingdao, 25 wanted criminals could be identified and arrested with the help of Megvii's advanced facial recognition software – admittedly a much better result than the London Metropolitan Police's use of CCTV during the Notting Hill Carnival of 2017 discussed in the previous part. The Economist (2017) reports that the software is now routinely used in Shenzhen "to identify jaywalkers; names and pictures go up on a screen [while in] Beijing the municipality has started using the technology to catch thieves of toilet paper in public restrooms." Furthermore, many Chinese hotels already use facial recognition software to verify the identity of a guest about to check in, and that railway stations like the one in Wuhan only admits passengers who are willing to have their face scanned, also for identification and verification purposes (ibid.) – further indicators of a rapid spread of this technology, unencumbered by human rights discussions. Interestingly, but probably not too surprisingly, at least one start-up firm, D-ID from Israel, already offers software geared to defeat such Smart CCTV systems by slightly altering "photos so that algorithms cannot recognise them" in order to be able to post and share photos without being recognized, while others recommend the use of simple means to fool surveillance systems, "such as glasses with hallucinogenic patterns on the frame of the specs, or simply wearing masks or make-up" (The Economist 2017).

That the Chinese government does not exactly start from scratch regarding a country-wide network of CCTV systems (known as *Skynet,* by the way, although it is unclear whether any irony is involved or not), is demonstrated very drastically by an experiment arranged by BBC News in December 2017. The question was simple: "How long can a BBC reporter stay hidden from CCTV cameras in China? The reporter in question was John Sudworth who, for the purpose of the experiment, had his photo taken and added to the state database. He then walked through the city centre of Guiyang towards the railway station (albeit without trying to hide from the CCTV systems he encountered). It turned out that it only took seven minutes until he found himself surrounded by police. Even the creators of the surveillance system behind it expressed a certain sense of discomfort when interviewed for the short BBC video clip on Sudworth's adventure (see at BBC 2017c). Another drastic example for how saturated with CCTV at least the urban areas of China already are, is the documentary-style mystery movie by Chinese artist Xu Bing, released in late 2017. The movie, titled *Dragonfly Eyes* (Qing Ting in the Mandarin original), makes do without a cameraperson – rather, it was produced simply by downloading from a selection of the already existing 170 million CCTV systems "hundreds of hours of publicly accessible surveillance videos online, many from Chinese livestreaming sites" (S. Chen 2017). The no less than 11,000 hours of surveillance videos thus collected were then categorized, edited into a storyline and turned into an 81-minutes film (Strittmatter 2018). Unsurprisingly, the movie has been dubbed a 'real-life Truman Show', after the famous 1998 US satirical movie. According to Xu Bing, the intention of his movie is to force "viewers to reflect on the 'surveillance state' people live in everywhere now" (ibid.). The movie's title was not chosen at random or for its poetic notion only – rather, it draws further attention on the already existing possibilities of surveillance via CCTV via the composition of a dragonfly eye, which according to Strittmatter (2018) consists of 28,000 facets that recognizes images five to six times faster than a human eye, and that it is also able to monitor 360 degrees of its environment all the time.

Smart CCTV systems however are by far not the only surveillance devices Xu Bing's viewers have to ponder – and we as well. Rather, and as I already mentioned in the previous part, it is very likely that in the near future, the all-seeing eyes of smart CCTV systems will be supplemented by equally all-seeing eyes in the shape of augmented reality glasses similar to Google's smart glasses. In China, such devices, equipped with state-of-the-art facial recognition software, are already in the process of being issued to police officers, for example in the city of Zhengzhou. These smart glasses do much of what the Smart CCTV systems can do: identifying individuals and adding name plus address to the picture. Unlike Megvii's sophisticated facial recognition software developed for the network of Smart CCTV systems powered by state-of-the-art computers, the smart glasses version still is less capable, as discussed in the previous part: after all, the officer needs to snap a picture of a suspect first before the identification process can begin. This notwithstanding, the effortless identification of suspects just by looking at people while on patrol no longer belongs to the realm of science fiction or movies such as *Minority Report* but to our reality – at the moment, and at least in China, rapid technological advance seems to be getting there rather quickly.

Seen from a civil liberties perspective, it is indeed a relief to know that, according to The Economist (2017), [the] West is further behind [and] it is mainly big online companies [such as Google, Facebook, Amazon, or JetBlue and Lloyds Bank] that make (cautious) use of the technology." In a sense, in the West, we are not there yet, and it still is up to us to decide whether we will get there – or, more realistically, when and under what conditions and circumstances.

13.2 The Gamification of Social Control: Towards 'Funopticon'

With the combined surveillance power of smart CCTV and smart glasses, we have however not yet reached the end of the 'slippery slope' from being a liberal democracy (our starting point) towards an authoritarian police state (where we could end) – we are just mid-way, so to speak. Above, I mentioned the possibility to establish exclusion zones or inclusion zones for convicted criminals required to wear electronic ankle tags. As I mused, most of us would probably be fine with this kind of civil liberties restriction since it would not affect 'us' as law-abiding citizens but only 'them' as rightfully convicted criminals. But what if we extend this monitoring to encompass not only a relatively small number of convicted criminals but to a much larger number of people suspected of being potential law-breakers, or potential terrorists, or at least potential sympathizers of terrorists, maybe due to their ethnic origin, or their religion? Most of us would baulk at the very idea, and rightfully so. Again, China is at the forefront of such developments, currently using villages in the Western parts of its restive Muslim-majority Xinjian province as testbeds for sophisticated facial recognition software paired with Smart CCTV systems. The villages selected serve as 'inclusion zones', with alarms being triggered if and when "targeted people venture more than 300 metres (1,000 feet) beyond designated 'safe areas' [which] comprise individuals' homes and workplaces" (Bloomberg 2018). Although in contrast to the Scottish prisoners described above, it is not strictly forbidden for the villagers to leave their safe areas, doing so might lead to some officials tracking their movements via the plethora of police checkpoints and CCTV systems deployed in this region. To facilitate this, inhabitants actually have to actively participate in the efforts to monitor their movements. For example, it is compulsory for them to equip their cars with GPS systems, or to undergo facial identification and verification checks whenever they "enter markets, buy fuel or visit places such as the capital Urumqi's main bus terminal" (ibid.). As one critic of this system puts it, "A system like this is obviously well-suited to controlling people […] 'Papers, please' was the symbol of living under a tyranny in the past. Now, government officials don't need to ask" (Jim Harper, as quoted in Bloomberg 2018). Again, and even if this is akin to labouring the point, it is good to know that we in the West are not there yet – but this second example from China shows what is already technically possible on the 24/7 surveillance front.

There is a third example that needs to be mentioned here that again demonstrates how far it is already possible to go due to modern security and counter-terrorism technology. Above, I already mentioned China's state database, questioning whether indeed about 1.3 billion Chinese (and thus, basically all Chinese citizens) are already part of it, or whether it is 'only' those 700 million who have been issued with an ID card. As Hatton agrees, such an all-encompassing national database that holds the information of each and every Chinese citizen is not yet in place – however, building up such a database is actually the plan, again with 2020 serving as the deadline: "By 2020, everyone in China will be enrolled in a vast national database that compiles fiscal and government information, including minor traffic violations, and distils it into a single number ranking each citizen" (Hatton 2015). What we are talking about here basically is a plan to assign each Chinese citizen a 'social score' based on a complex algorithm that can go up or down depending on their actions, and on how 'trustworthy' they are deemed to be in the eyes of the state (Botsman 2017). At the moment, there are several smaller pilot projects run by eight Chinese private sector companies (among them a matchmaking service and a financial service, Sesame Credit) testing the feasibility of the ambitious plan of a 'Social Credit System' or SCS for short. One spokesperson of the financial service is quite blunt on how his company assesses their customers:

> Someone who plays video games for 10 hours a day, for example, would be considered an idle person, and someone who frequently buys diapers would be considered as probably a parent, who on balance is more likely to have a sense of responsibility. (Li Yingyun of Sesame Credit, as interviewed in Hatton 2015)

If and when this national database goes online in 2020, all these available bits and pieces of information will be merged with data derived from government sources and, if possible, also with 'Big Data' as described in Part II above. For those ticking all the boxes, that means for all those who are on the right side of the state, this database would make their lives easier – for example checking in to a hotel without paying a deposit due to their exemplary credit score, or getting preferential seating in high speed trains or planes, allowing them to jump the queue in front of the ticket counter, ambling past all those who score lower than they do. And since this is quite a prestigious thing to do, Sesame Credit (one of the private sector companies currently developing the SCS) is actually encouraging its high-score users to flaunt their high-score – in the hope that others either feel encouraged to (or at least feel brow-beaten to) try to increase their own personal score as well (Hatton 2015). Botsman (2017) even notes that [higher] scores have already become a status symbol, with almost 100,000 people bragging about their scores on Weibo (the Chinese equivalent of Twitter) within months of launch." That the app has the look and feel of a cute and colourful gaming add instead of a dour credit scoring one certainly helps in that regard (Osborne 2015). In a sense, Huxley's *Brave New World* comes to mind (albeit he did not include such a database in his plot). However, woe betide those who for one reason or another end up on the wrong side of the state, as Simon Denyer points out:

> In this world, anything from defaulting on a loan to criticising the ruling party, from running a red light to failing to care for your parents properly, could cause you to lose points. And in this world, your score becomes the ultimate truth of who you are – determining whether you can borrow money, get your children into the best schools or travel abroad; whether you get a room in a fancy hotel, a seat in a top restaurant – or even just get a date. (Denyer 2016)

The official aim of this project is to create "a culture of 'sincerity' and a 'harmonious socialist society' were 'keeping trust is glorious", with the overarching principle being "If trust is broken in one place, restrictions are imposed everywhere" (Denyer 2016). The tools of this project can be described as a combination of carrot and stick: compliance results in easy access to coveted services such as described above; while non-compliance would not only involve the gradual loss of privileges such as "slower internet speeds, restricted access to restaurants and the removal of the right to travel" (Botsman 2017), but also in increased surveillance and even random checks on the unfortunate citizen who managed, for a wide variety of possible reasons, to lose the 'trust' of the state (ibid.). The lower one's social credit score, the more severe the restrictions – that is, in a nutshell, the message: try hard to be a model citizen, or else...

Within academia, this combination of carrots and sticks is known as 'nudging' – a way of social control deemed to work better than the resort to brute force only (Sunstein 2014). More to the point, given the context of high tech, is another term that gained currency in the meantime, and this is the 'gamification' of social control – that is, using the mechanics of computer games in the context of enhancing social control as tools to modify individual behaviour, as described in the case of China's social credit system (on gamification in general, see for example Dymek and Zackariasson 2017). Indeed, it seems that Randolph Lewis is quite right when he quips, as already mentioned above, that surveillance does not feel like a *Panopticon* any longer, but more like a *Funopticon* (Lewis 2017, 52).

13.3 'Total Information Awareness' and 'Undemocratic Means': Western Liberal Democracies and the Lure of Social Control

Nobody, not even its own government, would claim that the People's Republic of China is a liberal democracy, which implies that both socio-politically as well as geographically, it is far removed from our own Western liberal democracies, and of our open societies. So, why should it matter for us what the Chinese government does? As I already said, we are not there yet – neither technically nor mentally. But then again, maybe we should pause and think: what have we learned in the previous part about CCTV and Smart CCTV already deployed (or soon to be deployed) in our own cities? What have we learned about various biometrics, especially about facial recognition or gait recognition and their suitability for remote monitoring? What have we learned about Big Data, RFID and GPS-enabled devices and the Internet of Things – a technology that grows in leaps and bounds, and that will vastly increase our personal digital fingerprints? And what have we learned about huge databases such as Total Information Awareness? If we inspect and discuss all these issues individually, we might shrug and agree with Margaret Thatcher that sometimes, undemocratic means need to be employed to defeat a serious threat to our societies and, thus, our individual survival. But if we connect all the dots, and step back to take a look at the picture as a whole, we can see that we are not that far away from getting our own personal social credit score. In parts, such a score already exists in any case – credit reporting agencies such as Equifax or Experian are already routinely calculating such scores based on the available information on us. What distinguishes us from China is that so far, the state does not seem to be interested in connecting those dots for its own purposes, as the Chinese state does. And it is very much in our interest to keep it that way – if we are not carried away by the discourse of fear and a sheer technocratic, uncritical understanding of security, that is.

In order to understand just how far we already got carried away on the slippery slope from liberal democracy to something much less liberal, we only need to take another look at database programs such as the FBI's *Next Generation Identification* (NGI) or the US Department of Defense's *Total Information Awareness* (TIA). In the previous part, I already discussed them, but with a focus on their importance for the fight against terrorism. Now it is time to take a close look at their impact on our civil liberties. Quite a telling example for that is provided by Kyle Chakya, drawing on 'whistle-blower' Edward Snowden:

> From 2008 to 2010, as Edward Snowden has revealed, the National Security Agency (NSA) collaborated with the British Government Communications Headquarters (GCHQ) to intercept the webcam footage of 1.8 million Yahoo users. The agencies were analysing images that they downloaded from webcams and scanning them for known terrorists who might be using the service to communicate, matching faces from the footage to suspects with the help of a new technology called face recognition. (Chakya 2014)

In the previous part, I also described a German data-mining effort undertaken in the immediate aftermath of 9/11 with the aim to uncover sleeper cells of Al Qaeda-related terrorists hiding in Germany. I quoted Amoore and de Goede's laconic remark that the "search did not yield any terrorist suspect or prosecution" (Amoore and de Goede 2008, 7). As regards the NSA-GCHQ cooperation, Chakya is more descriptive, but equally dismissive:

> The outcome was pure Kafka, with innocent people being caught in the surveillance dragnet. In fact, in attempting to find faces, the Pentagon's Optic Nerve program recorded webcam sex by its unknowing targets – up to 11 per cent of the material the program collected was "undesirable nudity" that employees were warned not to access, according to documents. (Chakya 2014)

Chakya also quotes Jennifer Lynch from the Electronic Frontier Foundation (EFF). Lynch draws attention to the fact that the already mentioned Next Generation Identification (NGI) programme of the FBI does not differentiate between biometrics collected during criminal investigations and those collected from applicants for public sector jobs. This fusion of criminal and non-criminal data means that "[if] cameras connected to databases can do face recognition, it will become impossible to be anonymous in society" (Lynch, as quoted in Chakya 2014). In Chakya's view, this again means that "[in] the future, the government could know when you use your computer, which buildings you enter on a daily basis, where you shop and where you drive. It's the ultimate fulfilment of Big Brother paranoia" (Chakya 2014).

This 'Big Brother paranoia' already has a measurable impact on the behaviour of many users of social media: there is a noticeable trend towards self-censorship – a trend known as 'social cooling.' As the Dutch data protection activist Tijmen

Schep puts it, initial thoughts are no longer always acted upon: "I could click a link, but it might not look good. Somebody who is observing me could see it, hence I hesitate" (as quoted in Brühl and Tanriverdi 2017), translation mine). As Brühl and Tanriverdi observe,

> the knowledge that the data pools of social media, of sales portals and the surveillance tools of authorities could have repercussions for one's own life descends like frost on the societies. In a time of Big Data, this social pressure leads towards self-censorship as well as towards a reduced willingness to risk anything (Brühl and Tanriverdi 2017, translation mine).

In my opinion, this trend towards social cooling and self-censorship does not augur particularly well for the future of our liberal democracies – after all, such kind of behaviour is the hallmark of an authoritarian regime, as described by Orwell. Be that as it may, modern surveillance technology in conjunction with the laws paving the way for their use do not only have an impact on terrorists and their activities, but also a measurable one on the behaviour of average citizens.

At the time of writing, we are already one step further down the slippery slope from liberal democracies towards illiberal democracies – again due to the willingness to adopt, at least temporarily, undemocratic means in order to protect our democracies from the onslaught of transnational terrorists fighting for Al Qaeda or ISIS. Above, I mentioned the compulsory participation of villagers in China's Xinjiang province in their own observation. In case this sounds like a typical measure of an authoritarian state that would never been adopted by the governments of our liberal democracies, consider this: due to increased security measures at US borders, there is now a certain risk that travellers chosen (at random) for a secondary inspection are asked to unlock and hand over their smart phones to the immigration officer so that the contents of their phones can be checked, even without any suspicion that the traveller in question committed any crime, and without the traveller being on a watch list. For non-US citizens, refusing to comply and thus to basically bare their private lives could result in not being admitted – which means there is not much that can be done to prevent this blatant intrusion into one's privacy apart from leaving one's smart phone at home and to carry a cheap one without any potentially embarrassing material on it in order to avoid suspicion (Brühl 2018). After all, in our modern world, not having a smart phone does indeed look odd and suspicious.

In this context, and since in the first part, I mentioned the convenience of being enrolled in 'smart border' programmes that enable the traveller to bypass queues with (more or less) consummate ease, and also appropriated Matthew Sparke's concept of 'cushy cosmopolitanism' for that purpose, it is only fair to add that Sparke saw this in a rather negative light since in his point of view, 'voluntarily' enrolling in such a border-crossing programme comes with "an element of self-policing" (Sparke 2008, 147). Sparke also draws attention on a possible flip-side of such programmes, that is in the creation of what he calls "extraordinary others" (Sparke 2008, 149–154):

> One obvious underside to the transnational citizenship of expedited crossing lanes is the slowed down border crossing experience imposed on ordinary travellers who cannot afford to purchase or do not have the organizational capacity or the desire to acquire membership in the fast lane. In his analysis of the code spaces of contemporary airports, Adey suggests that in this way that there are also emerging kinetic underclasses moving alongside – but much more slowly than – the fast lane kinetic elites (Sparke 2008, 149, with reference to Adey 2004, 1376)

Personally, I think that this is a bit overdoing it – after all, such programmes are aimed at the frequent (business) travellers, not at the occasional border crossers in the shape for example of vacationers. Sparke however is not wrong to juxtapose this 'cushy cosmopolitanism' with a 'carceral cosmopolitanism' that affects all those 'others' or 'undesirables' who do not tick all the boxes of the security algorithms behind these programmes, such as the Department of Homeland Security's *Automated Targeting System* (ATS) discussed in the previous part:

> [these travellers] also sometimes ironically experience very rapid movement too: rapid movement into detention centres, and, ultimately, rapid transnational movement out of America, sometimes into incarceration elsewhere. (Sparke 2008, 149)

In their case, it is obvious that whole states can be turned into the 'exclusion zones' I mentioned above in the case of ankle-tagged Scottish prisoners – made possible by modern cutting-edge security technologies developed mainly to keep transnational terrorists at bay.

Finally, I argue that with regard to the loss of anonymity which is part and parcel of many types of enrolment programs, we also are already several steps down the slippery slope by now. For instance, and as already discussed in the previous part, Governor of New York Andrew Cuomo introduced an ambitious high-tech makeover in July 2016 for New York's bridges and tunnels that included the installation of CCTV systems complete ANPR capability and facial recognition software. From an anti-terrorism (or anti-crime) perspective, there is nothing wrong with that, while for the average user, such a system would indeed be very convenient as there would be no need any longer to stop at toll booths to pay one's fee since the system could automatically identify a driver, either by reading their face or their car's number plate. New York's Civil Liberties Union was quick to highlight the negative aspect:

> Governor Cuomo's plan has the potential to put thousands and thousands of people's images and data in a massive database that could be used by the government for who knows what. There is also an enormous risk that innocent people will be misidentified as terrorists, especially people of color who are more likely to be inaccurately identified by the technology. We are taking one step closer to the dystopian world of Minority Report without any discussion of the serious privacy concerns that are implicated. (as quoted in Moon 2016)

Regarding the enormous risk of misidentifying and then criminalizing perfectly innocent people, it is important to remember some arguments made in the previous part on the weaknesses of facial recognition software: many of the currently available solutions still seem to come with an inbuilt tendency to identify some subsets of the population easier than others – a tendency called 'identification bias.' As Introna and Wood explain, due to the propensity of current facial recognition systems to create false positives, the "bias group (African-Americans,

Asians, dark skinned persons and older people) may be subjected to disproportionate scrutiny, thereby creating a new type of 'digital divide'" (Introna and Wood 2004, 192). Of course, we are now entering the 2020s, and a lot of progress certainly has been made on the facial recognition front. Incidents such as the Notting Hill Carnival of 2017 however demonstrate that this 'digital divide' still seems to linger on – a divide that is nothing else than the 'othering' I already mentioned several times.

Such cases of misidentification and the follow-on outcry of civil liberty activists they result in notwithstanding, it is probably fair to say that unlike in the Chinese case, there is no hidden plan of the government to set up a 'Big Brother' state – after all, what happens in New York might not necessarily be replicated in other parts of the USA. Even within much smaller states such as the UK or Germany, the decision to introduce or not any of those security measures such as ANPR, Smart CCTV and the like is an uneven one, with one county or federal state (or city) taking such steps while other do not. For the governments in London or Berlin, getting anywhere near to a 'Big Brother' society, even if there were the political will to do so, would thus be a long one beset with many obstacles. But then again, think of Margaret Thatcher's dictum discussed above that in order to defend democracy, we sometimes have to resort to undemocratic means, and remember that many of us would be at ease with this idea – after all, democracy is not a recipe for suicide, as Justice Barak of Israel's Supreme Court quipped (as quoted in Ganor 2005, 166–167). Michael Sorkin makes a very pertinent point in this regard, alluding to the dialectics of terrorism and counter-terrorism. He argues as follows:

> It is possible that the only answer to persistent 'terror' is a police state. The two form a perfect symbiosis and it is easy to understand the utility of regular attacks to the authors of the US government's 'Patriot' Act, the breeders of sniffer dogs, and the private security firms that have become such a growth industry. To produce both legibility and intimidation, the whole panoptic repertoire of spatial and social control is deployed with little objection. (Sorkin 2004, 255)

Here, Malcolm Anderson's 'politics of the latest outrage' come back into play: in the immediate aftermath of a major terrorist incident, the people – usually represented by conservative mainstream media as their cheerleaders – predictably clamour for harsher security measures to be implemented. In such an atmosphere of fear prevalent after an attack, the 'discourse of fear' can be used against us, now not by the terrorists who arguably triggered it, but by the government of the day speeding additional security measures, or whole packages of them, through parliament with consummate ease, and without the usually lengthy and acrimonious debates. To push this symbiosis idea a little further still, it could be argued that while terrorists are now able to conduct jujutsu-like complex terrorism (Homer-Dixon 2009, 4), governments can carry out complementary acts of ju-jitsu-like opportunistic law-making by using our fears against us. And so can big corporation, as we have seen. This does not augur well for the future.

Chapter 14
Democracy Transfigured: The Dawn of the 'Umpire State'

Abstract In the previous chapter, I focussed on the state. However, answering the question on how modern security technology impacts on our liberal democracies requires a look at actors other than the state as well – actors that are per definition non-state actors but who, in some regard, act like quasi-state actors. In this chapter, I assess these private actors in a critical perspective to see how they influence our daily lives. My main argument, following James Madison, is that 'the state' as such is running the risk of being reduced to an 'umpire state' in the shape of a political actor who, as a primus inter pares (first among equals) mediates conflicts between private actors in the shape of powerful corporations but is no longer able to control their 'quasi-feudal' domains. But, since our ability to predict the future arguably is quite limited, I end this chapter with a couple of thought-provoking questions: could it be that this fear of many 'Big Brothers' instead of just one is as overblown as the fear of terrorism?

Keywords State actors · Private actors · Private companies · Umpire State · Outsourcing · Data-mining · 'Big brothers'

Answering the question on how, and how much, modern security technology impacts on our concept of citizenship, and on our liberal democracies at large, requires a look at actors other than the state as well – actors that are notionally, and per definition, non-state actors but who, in some regard, act like quasi-state actors, with a certain disregard for the nation state that hosts them. The actors in question include the big corporations that have been mentioned several times in different chapters of the book: Apple, Google, Facebook, Amazon for example on the 'techno giant' front, and others such as private security companies and intelligence providers that are less well-known to us but also have an impact on our lives and our privacy. The technological clout of these business ventures has been discussed several times as well, for example regarding the Internet of Things, and their ability to collect and collate huge amounts of data, including Big Data, that we voluntarily surrender to them without thinking. Now it is time to re-visit them and to assess them in a much more critical perspective to see how they influence our daily lives.

P. Lehr, *Counter-Terrorism Technologies*, Advanced Sciences and Technologies for Security Applications, https://doi.org/10.1007/978-3-319-90924-0_14

14.1 Big Brothers Are Watching You: Going Beyond Orwell

As I already pointed out, surveillance does not feel like suppression any longer but like fun (Lobe 2017) and is essentially gamified. The idea of surveillance as fun or even as a kind of a game, and the consequences of this kind of thinking, is reflected in a rather interesting comment of Scottish author and bookseller Shaun Bythell on George Orwell's famous novel *1984*:

> Much of this novel has proved prescient, although today it's not the state installing surveillance devices in our houses, it is us inviting them in to keep an eye on us. With [Amazon's] Alexa and [Google's] Home someone is watching you but it's not Big Brother, it's Amazon and Google, and we're paying to have them in our kitchens and sitting rooms (as quoted in Sunday Herald 2018).

It can indeed be argued that we voluntarily surrender our private data to these corporations because doing so benefits us, at least in general. Despite some glitches such as Alexa's well-reported tendency to break into a spooky and creepy laughter seemingly at random and even in the death of the night (Wong 2018), this and other devices related to the Internet of Things have the capability to greatly facilitate our lives. Even paying the price of getting tailor-made advertisement on one's social media apps, and maybe in the near future *Minority Report*-like in department stores and malls as soon as we enter could be seen as beneficial, or at the very least not as harmful. Nowadays, there are even apps that either encourage us to share our data in order to increase our credit rating like those discussed in the case of China, or that allow us to earn money with our data, such as DataWallet – all we need to do is giving these apps access to our social media accounts, and every other month, we might get a pay-out ranging between US$1 to US$50, depending on how active we are. Indeed, as Melia Robinson (2016) quips in her article on DataWallet, this "unusual service takes the 'if you can't beat 'em, join 'em' attitude to data-snooping." For those so inclined, surveillance is not only fun, it even makes money. So, could it not be that this problem is as overblown as the threat of terrorism?

The services offered to us via the Internet and the Internet of Things may well be benign and beneficial – I do not deny that, and I freely admit that I make use of social media as well with utter disregard for what that means for my traceability on the web or for what happens to my data (although I am probably also an example for 'social cooling' since I am now very careful about what to upload and what better not to upload). However, it is not only the relatively 'benign' techno-giants such as Google, Amazon, Facebook, Apple or Yahoo who are interested in what we do while we are doing it, and also not only state authorities such as NSA, CIA, GCHQ and others. Rather, there are private security providers as well as private risk assessment firms who are also interested in such things – and in their case, we might not be so easy-going regarding the involuntary sharing of our data. Actually, we should rather be alarmed when it comes to these companies and their interest in us. The private security and risk assessment industry in question here are profiting from an increasing trend known as outsourcing – outsourcing of core responsibilities of the welfare state that either can no longer afford to do that all by itself, or whose conservative government is of the opinion that 'market solutions' are always better than state solutions. Put in the context of the current outsourcing and privatisation of former key services offered by the state, this creeping empowerment of private security companies (also known as Private Security or Military Companies, PMSCs) as well as private intelligence providers is quite disturbing, and the implications of this process are far-reaching. Kaplan (2000, 81) already connected the possible dots on the basis of an often-overlooked fact:

> Of the world's largest economies, fifty-one are not countries but corporations. [...] The five hundred largest corporations account for 50 percent of world trade. Corporations are like the feudal domains that evolved into nation-states; they are nothing less than the vanguard of a new Darwinian organization of politics.

Writing 18 years later, Bill Bain offers a very similar argumentation. He states that "[tech] titans such as Apple and Google are already aware their societal influence now dwarfs antiquated systems of traditional government," also quipping regarding the question whether they pay their fair share of tax or not that "[tech] firms being able to dictate their own levels of tax perhaps suggest that they are *already* in charge" (Bain 2018b). Kaplan's and Bain's rather perceptive arguments also put the fear of an all-powerful state as envisaged by George Orwell in his famous novel *1984* to rest – albeit to a somewhat awkward rest: instead of one Big Brother, there are many little Big Brothers now that have the technical capability to keep us under observation 24/7 if they are so inclined. This is not the place to digress into a discussion of outsourcing and privatisation in any great detail – but since it does have an impact on our democracy, some comments on that are in order.

Regarding this outsourcing, we already discussed the Chinese start-up company Megvii that, like similar Chinese companies, develops the tools that the government in Beijing needs for their intended 24/7 surveillance of their citizens. Now it is perhaps a good idea to mention here that it is not only the People's Republic of China as an admittedly authoritarian state that outsources such sensitive work – rather, this is done in the West and in our liberal democracies as well and quite routinely. In the chapter on data-mining, I stated that the analytical core of *Multistate Anti-Terrorism Information Exchange* or MATRIX that ran between 2005 and 2006 was a software tool named *Factual Analysis Criminal Threat Solution* (FACTS). This tool was developed not by one of the US intelligence services themselves but by a private company named *Seisint* – one of many private companies developing data mining tools and software, used not only by state authorities but by private sector enterprises as well. Since we arguably have become 'digital persons' by now producing "rivulets of information [that] stream into electric brains to be sifted, sorted, rearranged, and combined in hundreds of different ways" (Solove 2004, 1) basically around the clock, we thus enable various interested parties to create what Solove calls 'electronic dossiers', that is electronic "collection[s] of detailed data about an individual" (ibid.). In the case of China social credit system, Rachel Botsman (2017) quipped "Big Data meets Big Brother" – this is, at least as concerns the sheer potential, not different here.

In the post-modern state of the near and probably dystopian future, these 'interested parties' most likely will be globally and regionally operating private corporations, including private security companies and equally private intelligence services. Since these industries tend to be rather opaque, it is unclear what their activities will amount to for the average citizen – at least for those who did not end up on certain watch lists due to faulty algorithms, as discussed in the previous part. What is clear however is that Bythell is quite right to argue that "someone is watching you but it's not Big Brother", but probably also quite wrong when he points at corporations like Amazon and Google in that context: it is rather the rising number of companies populating the private security and intelligence market – companies whose name we usually do not even now. Like 'Big Brothers' in their own right, they might enter in an awkward condominium with the state as such – which, after all, sheds many of its erstwhile core competencies as well as by gradually eroding its monopoly on the use of physical force as such. As Amoore and de Goede (2008, 13) argue, "[the] authority of private risk assessment firms, biometrics companies, homeland security consulting, financial data mining firms and so on suggests a significant realigning of public and private power in the war on terror." I certainly agree with that. As a result, in my opinion it is also fair to say that Orwell's famous 'Big Brother is watching you' now needs to be adapted and changed into 'Big Brothers are watching you.'

Furthermore, it can be argued that the activities of these actors will by definition be less benign than that of the liberal welfare state as we (still) know it: regarding the role of big companies as 'big brothers' in their own right, one should not forget that in a neo-liberal setting, the main objective of companies is maximizing their profits: after all, investors and shareholders demand to get a decent return for their investment. And this implies that even services that we take for granted and assume to be free will be turned into profit-making businesses – including first responders such as police, ambulance and fire-fighters. In a sense, and against the backdrop of already existing so-called 'premium services', we might end up in a 'pay as you panic' way, with privatized first responders coming to our rescue depending on the subscription we have: in due time if we are on basic, or abseiling from a helicopter within minutes after our emergence call if we are subscribed to (and have paid) the platinum service. At the moment, this still belongs to the realm of science fiction – especially the dystopian so-called *Cyberpunk* sub-genre, with William Gibson's brilliant debut novel *Neuromancer* at its heart. But, as Cowper and Buerger (2003, 8) point out, "[yesterday's] science fiction is rapidly becoming a very real part of everyday life."

It is however not necessarily the not-so-subtle 'pay as you go' approach transferred from the realm of mobile phone contracts to crucial first responder services that should concern us most, as problematic as this is for the future not only of our liberal democracies but of our welfare states. Rather, there are other and far more subtle tools that we should worry about. I already mentioned the targeted advertisement that we get, basically on a daily basis, as a 'price' for the otherwise free usage of social media apps such as Facebook, Facebook Messenger, Twitter and Instagram or search engines such as Google and Yahoo. Advertisement is, of course, an attempt to influence our behaviour by very subtly nudging us towards a certain product, whatever this may be, depending on our preferences and on the 'likes' we gave on Facebook, among others. But, as the case of Cambridge Analytica demonstrates, this kind of influencing our behaviour does not stop with trying to sell products and services – rather, it also includes attempts to influence our political behaviour, especially our voting preferences, by targeting us with "political messages tailored to the psychological traits of voters" (Timberg et al. 2018; also see Cadwalladr 2018). The former CEO of Cambridge Analytica, Alexander Nix, allegedly boasted in March 2018 that the company played a "pivotal role in securing US President Trump's election victory [because] we did all the research, all the data, all the analytics, all the targeting, we ran all the digital campaign, the television campaign and our data informed all the strategy" (Guyoncourt and Butterworth 2018). And he added an even more revealing remark that rings true in the times of so-called 'fake news': "Things don't necessarily need to be true as long as they are believed" (as quoted in Pratt 2018). This may well be an extreme case, but it is most certainly not the only one in which a well-tailored campaign on the basis of personal data of the users of sites such as Facebook might have influenced enough people to make a difference Gapper 2018). Hence, as Paul Bernal quips,

> Profiling, targeting, persuasive manipulation are the tools of the advertiser on steroids, provably effective and available to those with the money and intent to use them. Unless Facebook changes its entire business model, it will be used in ways that interfere with our lives – and in particular that interferes with our politics. What is more, it is only going to get *more* effective. (Bernal 2018, emphasis in original)

Arguably however, this very subtle attempt to influence our political behaviour via what its creator calls a "psychological warfare tool" (Cadwalladr 2018) has nothing much to do with counter-terrorism technologies as such, and with their impact on our ways of life. The case of Cambridge Analytica however illustrates how much the 'Big Brothers' of our times, and within our liberal democracies, can do, and the power they already wield over us. Hence, it is in order now to raise some general points regarding our liberal democracies and their future.

14.2 The Return of the Oligarchy: Private Companies Evolving into (Quasi) Nation-States

Again, we can refer to Kaplan to get us started. Musing about a possible return of oligarchy, he reminds us that "[modern] democracy exists within a thin band of social and economic conditions, which include flexible hierarchies that allow people to move up and down the ladder" (Kaplan 2000, 96). He follows this reminder with a stark warning:

> As new and intimidating forms of economic and social stratification appear in a world based increasingly on the ability to handle and analyse large quantities of information, a new politics might emerge for us, too – less the kind envisioned by progressive reformers and more like the pragmatic hybrid regimes that are bringing prosperity to developing countries. (ibid.)

Although the economic aspect he mentions does not seem to be of immediate relevance in the current context, the illustrations he offers are quite telling: he juxtaposes the comparatively well-off 'haves' from the far more numerous marginalized 'have-nots' in the shape (in his US-focused context) of "barely literate Mexicans on the U.S. border, working in dangerous, Dickensian conditions to produce our VCRs, jeans, toasters, earn less than fifty cents an hour, with no rights and benefits," to end with a very poignant question: "Is that Western democracy or ancient Greek-style oligarchy?" (Kaplan 2000, 97). He then reminds us of President Dwight D. Eisenhower's warning of a 'military-industrial complex', to add that it "may be equally wise to fear a high-tech military industrial complex" (ibid.). That his fears expressed nearly two decades ago were not far off the mark should be evident by now: the influence of high-tech companies such as Apple, Facebook, Google in the West (or rather global) has already been discussed, and so have Chinese start-up companies like Megvii. There is yet another stark warning that Kaplan (2000, 98) has in stock for us:

> If democracy, the crowning political achievement of the West, is gradually being transfigured, in part because of technology, then the West will suffer the same fate as earlier civilizations. [Instead of, as] we believe, [...] bringing freedom and a better life to the rest of humankind [we] are poised to transform ourselves into something perhaps quite different from what we imagine.

I contend that nearly two decades after Kaplan made these arguments, the shape of our future societies and their political systems is somewhat clearer now. One eminently important pointer is the still on-going process of privatisation – a privatisation that even includes services of the state that, until very recently, have been deemed to be 'core services', including those that had been the prerogative of the police and the military. The emergence of ever more powerful private sector actors such as corporations that now offer services that until recently used to be part of the state's monopoly of the legitimate use of physical force (according to Max Weber) implies that Kaplan (2000, 93) is also quite right to argue following James Madison that the modern state might devolve into "a mere 'umpire', refereeing among competing interests" – competing interests pursued by corporations he equates to "feudal domains that evolved into nation-states", opining that "they are nothing less than the vanguard of a new Darwinian organization of politics" (Kaplan 2000, 81). Again, against the backdrop of on-going and seemingly unstoppable process of privatisation, Kaplan's arguments made nearly two decades ago ring very true, and rather prescient at that. The same can be said for James Madison's argument as well – even though we no longer have to deal with 'feudal domains' but with private companies.

Again, we could look at Cyberpunk novels such as Neuromancer or Snow Crash, or movies such as *Blade Runner* to get an idea of how this would look like.

But actually, this is not necessary because in a sense, the post-modern dystopian state could look very similar to the early-modern state: there still would be a government, a people and a nation with more or less clearly delineated borders. However, the writ of the government would be limited and tempered by a number, or even a plethora, of other powerful political actors also active within the state's borders, but not necessarily confined to it. In the early-modern state, these actors usually had been feudal lords or military entrepreneurs such as the Italian *condottieri* of the Renaissance or Albrecht von Wallenstein during the Thirty Years' War (1618–1648), but sometimes mighty private sector companies as well, for example merchant associations such as the British and the Dutch East India Companies (EIC for the British company, and VOC for the Dutch one) ruling their own far-flung territories and maintaining their own armies and militias with a more or less polite disregard of what the central government wanted, and in the case of the EIC and the VOC also an utter disregard of what the colonialized subjects including their own rulers wanted.

14.3 Black Swans and Other Spoilers: Some Food for Thought

Having said that all, there are some warnings in order now. First of all, as I said in the introduction, our capacity for predicting the future is a rather limited one – even without 'black swans' interfering. Normally, when trying to predict the future, we tend to pick and choose some current and foreseeable trends, frame them in a way that makes sense to us, and then imagine that trend's continuation and development in a rather linear fashion. The framing that we do is guided by, or infused by, our own experiences, and our view of the world. If we mainly see opportunities, we will end up with a cheerful picture full of silver bullets and panaceas – not only regarding technology as a silver bullet against terrorism, but also technology as a silver bullet against poverty, crime, or climate change. If we see mainly risks in the seemingly relentless march of technology, we will end up painting a rather gloomy picture full of dystopias and 'big brother' states. To a certain extent, our assumptions shape the future – but not entirely so.

Secondly, and regarding the state as a 'big brother' snooping on us: The state, whether a 'big brother' or not, is not the only one who can snoop on us. It is also not only the 'big companies' who can do so (and certainly do to generate revenue). It can also be us doing the snooping (and stalking), or hackers – the latter ones with decidedly malicious intent. But since the rise of cybercrime is not the topic here, let us ignore this dimension and focus on 'us'. There are, as most of us know, a couple of apps available that allow us to snoop on our partner or someone else we are interested in for a variety of (not always honourable) reasons, easily concealable on the target's smart phone (to which we would need to have access however, assuming that we are no hackers). There are miniature cameras as well, hidden in pens or other gadgets that can be remotely controlled and thus allow us to snoop on somebody without his or her knowledge. Most of these gadgets are actually well-intended: think of baby monitors, for example. When I discussed the Internet of Things in the previous part, I quoted McEwen and Cassimally (2014, 1) to make sense out of it. They mentioned a teddy bear amongst the objects that can be equipped with a processor. One of the reason why we should wish to do so is the facilitation of monitoring our little ones without having to be physically present in their bedroom, simply by eavesdropping on them. Even better – at least for so-called 'helicopter parents': a number of children's smart watches made for kids between 5 and 12 years of age come pre-equipped with an app that allows parents to access the smart watch's microphone by remote and activate it without the kid noticing this, to listen in on the kid's conversation. Since the GPS signal of the smart watch reveals the kid's location, concerned parents also know at all times where the little one is. To put that bluntly, these apps allow the parents to spy on their children – just like other apps allow us to spy on our partners or neighbours. Ironically, the Internet of Things and the ever cheaper electronic surveillance gadgets available now seem to have 'democratized' surveillance and observation to quite an extent, including the 'harvesting' of data it we know how to do that. Apart from the cherished pass-time of stalking our exes on various social media apps, we now can police each other – oftentimes with the explicit encouragement of the police to report discovered misdemeanour, such as motor bikes using the pavement to bypass a traffic jam, or fly tippers. Obviously, the number of Big Brothers has proliferated by now in a way unforeseeable for Orwell.

Thirdly, and again regarding the state snooping on us, it is interesting to note that in the case of children's smart watches just mentioned, it was not a children's organisation who flagged that up first. Rather, it was Germany's federal telecoms regulator, i.e. a German federal authority, which raised the alarm because these apps make it possible to spy on anyone – as some parents apparently did, listening in on teachers during class, for example. Thus, in October 2017, the regulator recommended these devices (which it deemed to be 'unauthorized transmitting systems') to be destroyed. Earlier that year, the federal telecoms regulator also prohibited the sale of a doll known as 'My friend Cayla' since it also came equipped with an app that allowed the spying on children – which is why this device was deemed to be an unlawful 'espionage device' (Kaufman 2017). This may sound like an over-reaction by the proverbial 'nanny state.' However, if parents can listen in on their children without the little ones' knowledge, and also monitor where they are, well, so can hackers or other individuals with ill intent: most of the early Internet of Things gadgets communicated via unsecured Wi-Fi or Bluetooth connections, which made an intercept very easy. The main lesson here however is that it was the state itself that raised the alarm – how does that fit with the gloomy 'Big Brother' dystopias?

But even as regards private companies as 'Big Brothers' in their own right, it was good to see a push-back against the activities of Cambridge Analytica and Facebook's somewhat cavalier attitude to who can access all the data they hold on the billions of Facebook users. Tesla and Space X for example deleted their own Facebook accounts in protest, while many of Facebook's main advertisement partners such as

Mozilla or Commerzbank ended their cooperation with Facebook for the time being (Sommerlad 2018). Even more importantly for the avoidance of future large-scale data breaches, public and private sector representatives called for a regulation of social media services, in line with the existing regulation of conventional media services such as newspapers or the BBC. Again, how does that fit with the gloomy 'Big Brother' dystopias?

Chapter 15
Outlook: The Need for 'Critical' Critical Infrastructure Protection Studies

Abstract In this final 'outlook' chapter, I pull out three main issues that should be taken home as the main lessons of this book. Firstly, I draw attention on the importance of the art of critical infrastructure protection (CIP) as an effort to 'control the uncontrollable'. Secondly, I highlight the urgent need of a 'critical' critical infrastructure protection approach that steers away from the solely technocratic approach chosen so far. And thirdly, I conclude with the warning that a focus on counter-terrorism technology as a 'silver bullet' to the detriment of addressing the root causes of terrorism might lead to an 'arms race' between terrorists and counter-terrorists that cannot possibly be won especially not by us ordinary citizens who might well end up 'temporarily' surrendering civil rights in vain, and without getting them back any time soon since the war on terrorism is an interminable one.

Keywords Critical Infrastructure Protection (CIP) · Layered defence · Silver bullet · Coordinated terrorism · Fakeapp · Drone attacks · James Madison · Benjamin Franklin

So, what can we take home from the preceding pages and chapters? In the following concluding 'outlook' chapter, I shall refrain from repeating what I already said, and content myself with pulling out three main issues that should be taken home as the main lessons after having read this book. Firstly, the importance of the art of critical infrastructure protection (CIP) as an effort to 'control the uncontrollable'; secondly, the urgent need of a 'critical' critical infrastructure protection approach that steers away from the solely technocratic approach chosen so far; and thirdly, the warning that a focus on counter-terrorism technology as a 'silver bullet' to the detriment of addressing the root causes of terrorism might lead to an 'arms race' of sorts between terrorists and counter-terrorists that cannot possibly be won – especially not by us ordinary citizens who might well end up 'temporarily' surrendering civil rights in vain, and without getting them back any time soon since the war on terrorism is an interminable one.

P. Lehr, *Counter-Terrorism Technologies*, Advanced Sciences and Technologies for Security Applications, https://doi.org/10.1007/978-3-319-90924-0_15

15.1 'Feigning Control Over the Uncontrollable': Conventional CIP

The first issue that can be taken home is a rather general one: the effort to harden our open societies and our cities against terrorist attacks by marshalling the powers of modern high and low tech tools basically amounts to an attempt to "feign control over the uncontrollable", as Ulrich Beck sees it in a comment on 9/11, albeit putting 'risk' in a much broader context (2002, 41). In the narrower context of threats to security, especially terrorist threats to our security, then US Secretary of Defense Donald Rumsfeld can be quoted since he expressed the problems we face in this regard quite eloquently also against the backdrop of 9/11:

> The message is that there are no knowns. There are things that we know that we know. There are known unknowns. That is to say that there are things we now know we don't know. But there are also unknown unknowns – things we don't know we don't know. And if one looks throughout the history of our country and other free countries, it is the latter category that tend to be the difficult ones. (Rumsfeld 2002)

In the introduction and in the first part, I cautioned that even though the terrorist threat currently mainly emanates from Al Qaeda and ISIS, it would be good to keep an open mind regarding new manifestations of terrorism, with different sets of actors and objectives behind them. That a new generation, or a 'new wave' of terrorism (to echo Rapoport 2004) will sooner or later appear goes without saying – it is unlikely that Islamist terrorism will be the last we see of this type of political violence. However, we do not know what this new generation of terrorists will fight for. This, then, belongs to the 'known unknowns' according to Rumsfeld. These new terrorists, or just a newer generation of Al Qaeda or ISIS terrorists, may make use of new weapons and tactics to attack our societies 'ju-jitsu'-like as Homer Dixon (2009) predicts, thus taking us by surprise just as the 9/11 perpetrators did. This would belong to the 'unknown unknowns', and thus 'the most difficult ones' indeed, as Rumsfeld correctly points out. And this is exactly where critical infrastructure protection (CIP) in conjunction with modern security technologies comes in.

When I introduced CIP, I also mentioned in passing that this relatively new field of studies usually employs an 'all hazards' approach – that is a risk management plan valid for all kinds of risks, be they terrorist, criminal, human error, or natural disasters such as floods, earthquakes, blizzards etc. CIP thus is, by its very nature, not focussed on one specific threat scenario but tries to pre-empt and defeat any kind of threat, as former UK Prime Minister Tony Blair (2002) seemed to have acknowledged, knowingly or not, when he remarked that "[responsible] science and responsible policymaking operate on the precautionary principle." Indeed, the term 'precautionary' encapsulates the 'all hazards' approach of CIP risk management plans quite nicely. Since I mentioned Beck's much broader conception of risk, it is important here to repeat that within CIP risk management, risk can be defined as as "a function of consequences of a successful attack against an asset; and the likelihood of a successful attack against an asset" (API/NPRA 2004, 3; but see Aradau and van

Munster 2008, 29–33 for a different understanding). Hence, CIP risk management is not necessarily about eliminating the threat completely, thus reducing risk to 'zero', but to reduce the likelihood of a threat and to mitigate the consequences if a threat remains undetected.

With this caveat in mind, looking for a 'layered defence' that can protect or harden our cities against different kinds of terrorist attacks by harnessing high tech as suggested by Mills and Huber (2002), and low tech as well, makes eminent sense since it follows the precautionary principle. From a CIP perspective then, there is nothing wrong with biometrics, data-mining, scanners and sniffers, smart CCTV or citadels and barricades: all these measures work well within this 'layered defence' approach. Whether "our silicon" will eventually win as predicted by Mills and Huber or not is a rather moot point for risk managers – again, 'risk mitigation' is the objective for CIP, not necessarily 'risk elimination.' After all, Murphy's Law always seems to lurk in the background, ready to strike.

15.2 Ricocheting Silver Bullets: Arguments for a Critical CIP

The second issue that should be taken on board is a consequence of the first, and that is the realisation that technology is not necessarily a silver bullet as regards the fight against terrorism: apart from Murphy's Law, there is obviously some 'collateral damage' involved in using this silver bullet – collateral damage in the shape of a restructuring of our cities, and of our concept of citizenship as well as our understanding of democracy via an unquestioning, purely technocratic development and deployment of security technologies as discussed. Whether such collateral damage is a price worth to pay depends on one's conviction and probably also on the occurrence of particularly devastating attacks. After all, this is precisely what the 'politics of the latest outrage' and the 'politics of the last atrocity' imply: people usually clamour for strong and decisive action in the immediate aftermath of a terrorist event as the 'latest outrage' or 'last atrocity' but tend to lose interest when the outrage eventually dissipates. This implies that if the frequency of attacks is comparatively high, the outrage does not dissipate, and an atmosphere of fear prevails – an atmosphere conducive to a willingness to shrug off these 'collateral damages' that come with the silver bullet of technology, such as heightened surveillance or the (temporary) loss of certain civil liberties and human rights. In such an atmosphere, most citizens affected by acts of terrorism would probably subscribe to Margaret Thatcher's argument that "[in] order to protect democracy, you sometimes have no choice but to use undemocratic means", or even Shabtai Shavit's more pointed one that "[given] a choice between less democratic and surviving, and dying democratically, I prefer the first [option]" (as quoted in Ganor 2005, 166–167). On the other hand, if the frequency of attacks is not high enough to sustain the outrage, people probably would be more willing to listen to experts telling them that the modern London of the early twenty-first century is much more secure than the London of the nineteenth century, or that modern New York experiences far less violent crime than it did during the 1970s (Williams and Selle 2016, 19).

As Michael Gove famously quipped in the heat of the Brexit campaign, "Britain has had enough of experts" (Mance 2016). Being an expert myself, I think this rather sweeping and dismissive quip made out of political convenience rather than anything else misses the point. The issue – especially in our 'silver bullets' context – rather seems to be a certain disconnect between various sets of experts and those politicians and other stakeholders catering to the citizens' demands when it comes to the 'politics of the latest outrage.' Hugo Rifkind makes a very pertinent point in this regard when he assesses the repercussions of the Cambridge Analytica case mentioned above – the tailored political campaign that may or may not have swung the American elections of 2016. He argues as follows:

> We are here thanks to an unwitting alliance between politicians and pundits who didn't understand enough technology to grasp what was happening out there, and technological innovators who didn't understand enough politics to grasp why this was a problem. (Rifkind 2018)

On the basis of my experience, and the many talks I had with risk managers as well as technology providers, I entirely agree with Rifkind: there is not much information sharing and understanding between these two groups of people, with the result that a lot of possible information simply gets lost in translation, or falls prey to miscommunication. And this is exactly why I argue that a 'critical' critical infrastructure protection approach is required: it is simply not enough to churn out risk management plans based on available or soon-to-be-developed technologies without a critical appraisal of their impact on our open societies and our liberal democracies. Doing so betrays a tacit acceptance of Margaret Thatcher's position regarding the resort to undemocratic means in order to protect democracy, whether it is obvious or not. As discussed in this part, there might be a hefty price to be paid for this quest for 'zero risk', and the hope that 'our silicon will win.' Sometimes, the cure (of changing our societies and curtailing our civil liberties) might indeed be worse than the disease (of terrorism).

There is a final issue that needs to be flagged up: our quest for a 'silver bullet' in the shape of advanced technology harnessed to combat, and maybe even defeat, terrorism might result in the abandonment of the quest for an understanding of 'what makes them tick' – i.e. an understanding of the root causes of terrorism. Even nowadays, many hardliners among counter-terrorism professions, including academic ones, forcefully argue in favour of such attempts to be abandoned, usually either with the claim that trying to understand terrorists and terrorism would lead to mere apologetics, or at least with the assertion that such an attempt would be futile since identifying root causes admittedly is a rather difficult task, and most certainly one that is akin to 'mission impossible': even if some root causes can be found, it is not always possible to do something about them. Ending 'relative deprivation' for example would necessitate to rewrite the socio-economic foundations of society. Hence, concentrating on 'kinetic operations' abroad while hardening our societies, our cities and our critical infrastructures at the home front via the proper technological solutions sounds like a perfect strategy.

Unsurprisingly, as Sloan (2007, 20) notes, specialized counter-terrorism programmes "tend to favour technological innovation in regards to preventing or countering terrorism." He follows that "[unquestionably] the technological threats are there, but does this approach rely too much on scientific and technological fixes to address the at once complex and yet human intensive terrorist threats?" And finally, he asks a very pertinent question: "In the organisation of academic research on terrorism, are we running a similar risk of becoming technologically muscle-bound against an emotional and often value- or religious-driven, highly committed adversary?" His evocative picture of a technologically muscle-bound counter-terrorism warrior that reminds me a bit of the comics and Hollywood blockbuster character Judge Dredd brings me back to the 'agency of things' idea discussed in the introduction: since all these gadgets that we use have the tendency of gradually changing us and the way we do things, what impact then have the counter-terrorism technologies just discussed on us and our societies? What are these technologies doing 'to us', apart from 'for us'? And what about the terrorists themselves? Will they also morph into equally technologically muscle-bound experts in coordinated terrorism, bringing our societies to their knees in a jujitsu-like use of the very same high-tech against us, just as Homer-Dixon (2009) speculated?

15.3 'Muscle-Bound Experts in Coordinated Terrorism': An Avoidable Future?

The newest fear regarding high-tech terrorism is the fear that well-equipped and well-trained high-tech terrorists could use very similar computer systems based on Artificial Intelligence (AI) that we deploy against them for postdictive or even predictive data mining against us – not necessarily to conduct some 'counter-datamining' operations but to launch sophisticated attacks previously out of their reach. For example, they could use AI to defeat security systems based on seemingly (more or less) unique personal traits in order to gain access. A recent report on the malicious use of AI describes such a scenario:

> For example, most people are not capable of mimicking others' voices realistically or manually creating audio files that resemble recordings of human speech. However, there has recently been significant progress in developing speech synthesis systems that learn to imitate individuals' voices [...] There is no obvious reason why the outputs of these systems could not become indistinguishable from genuine recordings, in the absence of specially designed authentication measures. Such systems would in turn open up new methods of spreading disinformation and impersonating others. (Eckersley et al. 2018, 20)

This scenario is only a hypothetical one – at least so far. Our voices however are not the only biometrical trait that could (and at some time will be) used against us by terrorists with the help of AI systems. The next logical biometric, also widely used for identification and verification purposes, is our face. One example on how this could be used against us as well is a freely available app named *FakeApp* which can be downloaded free of cost to our device of choice. What it does can be described simply as 'swapping faces'. *FakeApp* draws on Google's *TensorFlow* as an "open-source machine learning framework for everybody" (tensorflow.org, undated) for that purpose. As Marino (2018) elaborates, "FakeApp works by using thousands of images and videos of the person's face a user is trying to fake, in order to build an AI generated model that can then be applied to a video of the user's choice." Such edited videos are termed *DeepFakes* and, as Marino also explains, some of them are created for comedic purposes, for example by replacing the characters of popular movies with stars such as Nicholas Cage, or replacing the face of Dr. Evil from the famous Austin Powers movies with the face of current US President Donald Trump (ibid.). The results are convincing and realistic, although the process of creating even just a five-minute video "can take hours if not days" (ibid.). But sheer computing power is the only obstacle, as Bill Bain (2018a) observes: "Anyone who can work an app will be able to place you into any footage, artificial or otherwise. The fake you will be indistinguishable from reality." Swapping Dr. Evil's face for that of President Trump already shows how this app can be misused by interested parties, for example by creating fake 'celebrity porn' for blackmail, fake interviews with politicians for the purpose of disinformation – or for fooling identification and verification systems based on remote facial recognition such as Smart CCTV systems. Unsurprisingly, Eckersley and his co-authors warn in the executive summary of their report that

> [new] attacks may arise through the use of AI systems to complete tasks that would be otherwise impractical for humans. In addition, malicious actors may exploit the vulnerabilities of AI systems deployed by defenders (Eckersley et al. 2018, 5).

If the scenario of AI-supported high-tech terrorists engaged in a lethal match of ju-jitsu with counter-terrorism forces sounds a tad too outlandish to us (admittedly, such a scenario still is in the future), then we could settle for yet another frightening one: terrorists using drones against us. This has already been done, occasionally – but the increasing availability of comparatively cheap off-the-shelf drones that can carry a substantial payload makes it even more likely to happen in the near future. The drone in question does not even need to be loaded with explosives: steering it in the landing path of an airliner could be all that is required to bring an airport at least temporarily to its knees: even if the airliner so targeted makes a successful emergency landing, starts and landings would be suspended for hours until the police would give the all-clear. There have been many reports of drones straying into the landing paths of airliners, but so far probably out of error, and without causing a collision. But it is probably only a question of time until some terrorists, possibly self-starters without any connection to experienced bomb makers of Al Qaeda or ISIS might plan such an attack.

But I don't intend to digress into an exercise of 'red-teaming' here by thinking like a terrorist and conjuring up probable (and improbable) attack scenarios. My point is a different one, closely following Sloan in this regard: by focussing too much on our silicon as Mills and Huber (2002) put it, we might force terrorists to do likewise. We might force them into an arms race, sort of – after all, terrorists always look for chinks in our armour, and sooner or later, they will find some. This also implies that we might give up some of our civil liberties, or parts of our way of life, in vain. We might find out to our peril, and belatedly, that Margaret Thatcher's argument that sometimes, we need to use undemocratic means to defend our democracies may well sound persuasive, but that it is wrong nevertheless. And with that, it is time to quote two of the Founding Fathers of the USA, James Madison and Benjamin Franklin. Madison's remark is rather pertinent against the backdrop of our quest to harden our societies against a mainly foreign threat – the one posed by Al Qaeda and ISIS:

> Perhaps it is a universal truth that the loss of liberty at home is to be charged to provisions against danger, real or pretended, from abroad. (Madison 1798)

Regarding the second quote, Benjamin Franklin's famous warning can be seen as a rather stern and succinct retort to Margaret Thatcher, Justice Barak, and Shabtai Shavit as well as other likeminded individuals, politicians, counter-terrorism specialists and ordinary citizens alike:

> Those who would give up essential Liberty, to purchase a little temporary Safety, deserve neither Liberty nor Safety. (Franklin 1755)

Of course, we simply do not know what Madison and Franklin would have said in the aftermath, and under the impression, of 9/11. Maybe they would have reiterated their positions, or they would have adapted them. Even so, their warnings are quite telling, and eminently relevant for our times as well. Again, whether collateral damage in the shape of living somewhat less democratic but at least surviving is a price worth to pay depends on our personal convictions, and probably also on the occurrence of particularly devastating attacks. I leave this to the readers to decide. In my opinion, it is a worthwhile effort to make best use of technology, both low tech and high tech, for our fight against terrorism. There is nothing wrong with that. However, there is also nothing wrong with adopting a more critical approach – not necessarily a 'technophobe' one, but rather a critical technophile one. After all, like 'real' bullets, silver bullets have the tendency to ricochet, as we have seen.

General Bibliography

9/11 Report Commission (2004) The 9/11 Commission Report: Final Report of the National Commission on Terrorist Attacks upon the United States (9/11 Report). Washington, DC.: Executive Agency Publications. https://www.gpo.gov/fdsys/pkg/GPO-911REPORT/content-detail.html

Acharya A, Mandal S, Mehta A (2009) Terrorist attacks in Mumbai: picking up the pieces. In: ICPVTR Report. Nanyang Technological University, Singapore

ACLU [American Civil Liberties Union] (n.d.) Surveillance under the patriot act. ACLU issues: National security. https://www.aclu.org/issues/national-security/privacy-and-surveillance/surveillance-under-patriot-act

Adey P (2004) Surveillance at the airport: surveilling mobility/mobilising surveillance. Environ Plan 36(8):1365–1380 http://journals.sagepub.com/doi/abs/10.1068/a36159

Agbola T (1997) Architecture of fear: urban design and construction response to urban violence in Lagos, Nigeria. Institut Français de Recherche en Afrique (IFRA), Nairobi

Al-Abed M, Charrier C, Rosenberger C (2012) Evaluation of biometric systems. In: Yang J (ed) New trends and developments in biometrics. INTECH, pp 149–169. https://www.intechopen.com/books/new-trends-and-developments-in-biometrics/evaluation-of-biometric-systems

Albanesius, C (2012) Google 'project glass' replaces the smartphone with glasses. PC Magazine, 4 April. http://uk.pcmag.com/consumer-electronics-reviews-ratings/64927/news/google-project-glass-replaces-the-smartphone-with-glasses

Allen J, Massey D, Pryke M (eds) (1999) Unsettling cities. Routledge, London/New York

Amoore L, de Goede M (2008) Introduction: governing by risk in the war on terror. In: Amoore L, de Goode M (eds) Risk and the war on terror. Routledge, London/New York, pp 5–19

Anderson M (2008) Counterterrorism as an objective of European police cooperation. In: Reinares F (ed) European democracies against terrorism. Governmental policies and intergovernmental cooperation. Ashgate, Aldershot, pp 227–243

Anderson B (2013) Us and them? The dangerous politics of immigration control. Oxford University Press, Oxford

API/NPRA (2004) Security vulnerability assessment methodology for the petroleum and petrochemical industries. American Petroleum Institute (API) and National Petrochemical & Refiners Association (NPRA), Washington, DC at http://www.npra.org/docs/publications/newsletters/SVA_2nd_edition.pdf

Aradau C, van Munster R (2008) Taming the future: the *dispositif* of risk in the war on terror. In: Amoore L, de Goode M (eds) Risk and the war on terror. Routledge, London/New York, pp 23–40

Argomaniz J, Lehr P (2016) Political resilience and EU responses to aviation terrorism. Studies in Conflict and Terrorism 39(4):363–379

P. Lehr, *Counter-Terrorism Technologies*, Advanced Sciences and Technologies for Security Applications, https://doi.org/10.1007/978-3-319-90924-0

Arnett P (1968) Major describes move. New York Times, 8 February

Asthana A (2016) Revealed: British councils used Ripa to secretly spy on public. The Guardian, 25 December. https://www.theguardian.com/world/2016/dec/25/british-councils-used-investigatory-powers-ripa-to-secretly-spy-on-public

Aust S (2008) The Baader-Meinhof complex. The Bodley Head, London

Avant D (2004) The privatization of security and change in the control of the force. Int Stud Perspect 5:153–157

Azad S and Gupta A (2011) A quantitative assessment on 26/11 Mumbai attack using social network analysis. Contemporary voices, 30 October. https://cvir.st-andrews.ac.uk/articles/10.15664/jtr.187/

Bača M, Grd P, Fotak T (2012) Basic principles and trends in hand geometry and hand shape biometrics. In: Yang J (ed) New trends and developments in biometrics, InTech, pp 77–99. https://doi.org/10.5772/51912https://www.intechopen.com/books/new-trends-and-developments-in-biometrics/basic-principles-and-trends-in-hand-geometry-and-hand-shape-biometrics

Bagnall N (2004) The Peloponnesian war: Athens, Sparta and the struggle for Greece. Pimlico, London

Bain B (2018a) Digital cloning is trump card of an artificial age. Sunday Herald, 04 March

Bain B (2018b) War? We've an app for that. Sunday Herald, 04 March

Barnes T (2018) Apple to move iCloud storage to China, prompting fears over government access to personal data. The Independent, 24 February. https://www.independent.co.uk/life-style/gadgets-and-tech/news/apple-moves-icloud-data-china-human-rights-spying-arrests-technology-privacy-a8226541.html?utm_term=Autofeed&utm_campaign=Echobox&utm_medium=Social&utm_source=Facebook

Baumann Z, Lyon D (2013) Liquid Surveillance: A Conversation. Polity Press, Cambridge/Malden, MA

Baumann I, Stephan A (2011) Kommissar Computer: Dr. Horst Herold (*1923) und die Geister, die er rief. In: Imanuel Baumann et al. (ed), Schatten der Vergangenheit. Das BKA und seine Gründungsgeneration in der frühen Bundesrepublik. Köln, Luchterhand, pp 79–86

Bay A-H, Blekesaune M (2002) Youth, unemployment and political marginalisation. Int J Soc Welf 11(2):132–139 https://onlinelibrary.wiley.com/doi/abs/10.1111/1468-2397.00207

BBC (2007) Car bombings suspect 'on the run'. BBC News, 2 July. http://news.bbc.co.uk/1/hi/uk/6259924.stm

BBC (2008) Soldier kidnap plotter given life. BBC News, 18 February. http://news.bbc.co.uk/1/hi/uk/7250697.stm

BBC (2013) Woolwich attack: lee Rigby named as victim. BBC News, 23 May. http://www.bbc.co.uk/news/uk-22644857

BBC (2016) French church attack: what we know. BBC Europe, 28 July. http://www.bbc.co.uk/news/world-europe-36900761

BBC (2017a) Groups call for Notting Hill carnival face scans to be dropped by met. BBC News England, 17 August. http://www.bbc.co.uk/news/uk-england-london-40963126

BBC (2017b) London attack: what we know so far. BBC News, 12 June. http://www.bbc.co.uk/news/uk-england-london-40147164

BBC (2017c) In your face: China's all-seeing state (video clip). BBC News China, 10 December. http://www.bbc.co.uk/news/av/world-asia-china-42248056/in-your-face-china-s-all-seeing-state

Beanland C (2018) Where the streets have no shame. The i Paper, 9 March

Beck U (2002) The terrorist threat: world risk society revisited. Theory, Culture & Society 19(4):39–55 http://journals.sagepub.com/doi/abs/10.1177/0263276402019004003

Beiner R (ed) (1995) Theorizing citizenship. State University of New York Press, New York

Bernal P (2018) The Cambridge Analytica scandal isn't a scandal: this is how Facebook works. The Independent, 19 March. https://www.independent.co.uk/voices/camridge-analytica-scandal-how-facebook-works-harvesting-data-politics-trump-brexit-a8264051.html?utm_campaign=Echobox&utm_medium=Social&utm_source=Facebook

Berner E (1998) Defending a place in the City: localities and the struggle for urban land in metro manila. Ateneo de Manila University Press, Quezon City

BI2 Technologies (2009) MORIS. Mobile, wireless and hand held offender recognition and information system. In: BI2 Technologies. http://www.bi2technologies.com/sites/default/files/documents/MORIS-Brochure.pdf

Biermann K (2015) BND stores 220 million telephone data – every day. Zeit Online, 2 February. http://www.zeit.de/digital/datenschutz/2015-02/bnd-nsa-mass-surveillance/komplettansicht

Biernat O (2015) Spitzentreffen zu 'No-Go-Areas' in Gelsenkirchen. WDR Nachrichten, 15 December, at http://www1.wdr.de/nachrichten/ruhrgebiet/spitzentreffen-zu-libanesischen-grossfamilien-100.html

Birrell I (2016) Nice attacker killed for reasons other than a twisted devotion to religion. I news, 17 July, at https://inews.co.uk/opinion/columnists/isis-red-herring-bouhlel-killed-reasons-twisted-devotion-religion/

Biswas AK, Tortajada C, Stavenhagen M (2018) Managing shrinking cities in an expanding world. The Conversation 22 February. http://theconversation.com/managing-shrinking-cities-in-an-expandingworld-91448

Bjørgo T (2005) Introduction. In: Bjørgo T (ed) Root causes of terrorism. Routledge, London, pp 1–14

Blair T (2002) Science matters. Speech delivered to the Royal Society, 23 May 2002. http://www.ukpol.co.uk/tony-blair-2002-science-matters-speech/

Blamont M (2016) Paris beach festival will be fenced in after nice attack. i Paper/Reuters, 20 July

Bloomberg (2018) China uses facial recognition to fence in villagers in far west. Bloomberg News, 17 January. https://www.bloomberg.com/news/articles/2018-01-17/china-said-to-test-facial-recognition-fence-in-muslim-heavy-area

Blount MD (ed) (2010) Police and augmented reality. Nova Science Publishers, New York

Blunden M (2018) Big brother is wondering why you're in the loo again. i Paper, 01 February

BMI (2016) Terrorismusbekämpfung. In: *Sicherheit,* Bundesministerium des Innern at http://www.bmi.bund.de/DE/Themen/Sicherheit/Terrorismusbekaempfung/Terrorismus/terrorismus_node.html

Booth R et al(2017) Tube attack: police search three properties as terror threat downgraded. The Guardian, 18 September. https://www.theguardian.com/uk-news/2017/sep/17/parsons-green-tube-bomb-police-arrest-second-man

Botsman R (2017) Big data meets big brother as China moves to rate its citizens. Wired, 21 October. http://www.wired.co.uk/article/chinese-government-social-credit-score-privacy-invasion

Boussadia K (2009) The evolution of airport screening technology. Biometric Technology Today Feature, February, 1–2

Boyd JE, Little JJ (2005) Biometric gait recognition. In: Tistarelli M, Bigun J, Grosso E (eds) Biometrics school 2003. Springer Verlag, Berlin/Heidelberg, pp 19–42

Brackett DW (1996) Holy terror. Armageddon in Tokyo. Weatherhill, New York

Bradsher K (2002) High and mighty: SUVs – the most dangerous vehicles and how they got that way. Public Affairs, New York

British Newspaper Archive (2013) Terror in the underground – 30 October 1883. The British Newspaper Archive Blog, 29 October. https://blog.britishnewspaperarchive.co.uk/2013/10/29/terror-in-the-underground-30-october-1883/

Britzelmeier E (2018) Wenn die Sicherheitskontrolle krank macht. Sueddeutsche Zeitung, 6 February. http://www.sueddeutsche.de/muenchen/flughafen-muenchen-wenn-die-sicherheitskontrolle-krank-macht-1.3853755

Brühl J (2018) Ist das ein Befehl oder eine Bitte? *Sueddeutsche Zeitung,* 14 March. http://www.sueddeutsche.de/digital/einreise-in-die-usa-ist-das-eine-bitte-oder-ein-befehl-1.3904502

Brühl J, Tanriverdi H (2017) Der überwachte Mensch zensiert sich selbst. Sueddeutsche Zeitung, 30 December. http://www.sueddeutsche.de/digital/kongress-des-chaos-computer-club-der-ueberwachte-mensch-zensiert-sich-selbst-1.3808605

Burgess M (2017) What is the IP act and how will it affect you? Wired, 8 May. http://www.wired.co.uk/article/ip-bill-law-details-passed
Burghardt T (2009) Curbing social dissent: 'non-lethal' weapons to be used in 'homeland security'? GlobalResearch, 6 September. https://www.globalresearch.ca/curbing-social-dissent-non-lethal-weapons-to-be-used-in-homeland-security/15078
Burton AM (1975) Urban terrorism: theory, practice & response. Leo Cooper, London
Cadwalladr C (2018) The Cambridge Analytica files: 'I made Steve Bannon's psychological warfare tool': meet the data war whistleblower. The Guardian, 18 March. https://www.theguardian.com/news/2018/mar/17/data-war-whistleblower-christopher-wylie-faceook-nix-bannon-trump
Caldeira TPR (2000) City of walls: crime, segregation, and citizenship in São Paulo. University of California Press, Berkeley
Calder S (2016) The temporary travel ban that's lasted 10 years. i Newspaper, 11 August
Carlo S (2016) The Snooper's charter passed into law this week – say goodbye to your privacy. The Independent, 19 November, at www.independent.co.uk/voices/snoopers-charter-theresa-may-online-privacy-investigatory-powers-act-a7426461.html
Caygill SJ, Davis F, Higson SPJ (2012) Current trends in explosive detection techniques. Talanta, Vol. 88, 15 January, 14–29. https://dspace.lib.cranfield.ac.uk/bitstream/1826/6887/1/Current_trends_in_explosives_detection_techniques-2012.pdf
Caygill SJ, Davis F, Higson SPJ (2012) Current trends in explosive detection techniques. Talanta 88:14–29 https://dspace.lib.cranfield.ac.uk/bitstream/1826/6887/1/Current_trends_in_explosives_detection_techniques-2012.pdf
Cesari J (2005) Ethnicity, Islam, and les banlieues: confusing the issues. Riots in France, SSRC.org, 30 November. http://riotsfrance.ssrc.org/Cesari/
Chakya K (2014) Face-recognition software: is this the end of anonymity for all of us? The Independent, 23 April. http://www.independent.co.uk/life-style/gadgets-and-tech/features/face-recognition-software-is-this-the-end-of-anonymity-for-all-of-us-9278697.html
Charlton A (2017) Police say Microsoft HoloLens and AR are a crime-fighting 'game changer'. GearBrain, 17 October. https://www.gearbrain.com/microsoft-hololens-ar-police-tool-2497703988.html
Chattopadhyay A, Chang CH, Yu H (eds) (2017) Emerging Technology and Architecture for Big-data Analytics. Springer, Cham/Singapore
Chauzal G et al (2015) Paris: 11/13/15. Analysis and policy options. Policy brief. Clingendael Netherlands Institute of International Relations and International Centre for Counter-Terrorism (ICCT), The Hague. November. https://www.clingendael.org/sites/default/files/pdfs/Policy_Brief_Clingendael_ICCTParis111315Analysis_and_Policy_Options_November%202015_final.pdf
Chen H (2006) Intelligence and security informatics for international security. Information sharing and data mining. Springer, New York
Chen LY (2017) China, Russia put millions in this startup to recognize your face. Bloomberg, 31 October. https://www.bloomberg.com/news/articles/2017-10-31/chinese-facial-recognition-startup-is-said-to-raise-460-million
Chen S (2017) This film is made entirely with surveillance footage. Quartz, 30 October. https://qz.com/1114178/chinese-artist-xu-bing-created-dragonfly-eyes-a-film-made-entirely-with-surveillance-footage/
Chertoff M (2006) A tool we need to stop the next airliner plot. Washington Post, 29 August
Christl W (2014) Kommerzielle digitale Überwachung im Alltag. Erfassung, Verknüpfung und Verwertung persönlicher Daten im Zeitalter von Big Data: Internationale Trends, Risiken und Herausforderungen anhand ausgewählter Problemfelder und Beispiele. *Studie im Auftrag der Bundesarbeitskammer Wien.* Wien: Cracked Labs. Institut für Kritische Digitale Kultur, November. http://crackedlabs.org/dl/Studie_Digitale_Ueberwachung.pdf
City of London Police (n.d.) Taser. City of London Police: About Us.https://www.cityoflondon.police.uk/about-us/your-right-to-information/Pages/taser.aspx

Clark RA (1988) Information technology and Dataveillance. Commun ACM 31(5):498–512 https://dl.acm.org/citation.cfm?id=42413

Clark M (2013 Google Glass Violates Nevada Law, Says Caesars Palace. IGN, 8 May. http://uk.ign.com/articles/2013/05/08/google-glass-violates-nevada-law-says-caesars-palace

Clarke AC (1973) Profiles of the future: an inquiry into the limits of the possible. Harper & Row (rev. ed.), New York

Clinton B (2002) Address by former US President Bill Clinton to the Democratic Leadership Council held at New York University, New York on Tuesday, 3 December 2002. www.amicc.org/docs/ClintonDec3_02.pdf

Coaffee J (2003) Terrorism, risk and the City. The making of a contemporary urban landscape. Ashgate, Aldershot

Coaffee J (2004) Recasting the 'ring of steel': designing out terrorism in the City of London? In: Graham S (ed) Cities, war, and terrorism. Towards an urban geopolitics. Blackwell, Malden, MA/Oxford, pp 276–296

Collins PA, Baggett RK (2009) Homeland security and critical infrastructure protection. Praeger Security International, Westport, CT/London

Cowper TJ, Buerger ME (2003) Improving our view of the world: police and augmented reality technology. Department of Justice/FBI, Washington, DC. https://www.fbi.gov/filerepository/stats-services-publications-police-augmented-reality-technology-pdf/view

Crabtree A et al (2013) A qualitative study of the perceived effects of blue lights in washrooms on people who use injection drugs. Harm Reduction Journal 10:22 https://www.ncbi.nlm.nih.gov/pmc/articles/PMC3853159/

Creekmore AT (2014) The social production of space in third-millennium cities of upper Mesopotamia. In: Creekmore AT III, Fisher KD (eds) Making ancient cities*: Space and Place in Early Urban Society*. Cambridge University Press, New York, pp 32–73

Creekmore AT, Fisher KD (eds) (2014) Making ancient cities: space and place in early Urban Society. Cambridge University Press, New York

Crelinsten R (2009) Counterterrorism. Polity Press, Cambridge/Malden, MA

Czeglédy AP (2003) Villas of the Highveld. A cultural perspective on Johannesburg and its 'northern suburbs. In: Tomlinson R et al (eds) Emerging Johannesburg: perspectives on the Postapartheid City. Routledge, New York, pp 21–42

Davis M (2002) Dead cities and other Tales. The New Press, New York

Davis M (2006a) *City of Quartz: Excavating the Future in Los Angeles*. London and. Verso, New York

Davis M (2006b) Planet of slums. Verso, London/New York

Davis M (2007) Buda's wagon. A brief history of the Car bomb. Verso, London/New York

Debray R (1967) Revolution in the revolution. MR Press

Decker RJ (2001) Homeland Security: a risk management approach can guide preparedness efforts, Testimony before the senate committee on government affairs, October 31. http://www.gao.gov/new.items/d02208t.pdf

Deen A (2016) I used to be an Islamic extremist. this is the truth about how you can prevent further terror attacks," The Independent, 20 July, at http://www.independent.co.uk/voices/i-used-to-be-an-islamic-extremist-this-is-the-truth-about-how-you-can-prevent-further-terror-attacks-a7145816.html

Deflem M (2005) 'Wild beasts without nationality'. The uncertain origins of Interpol, 1898–1910. In: Reichel P (ed) Handbook of transnational crime and justice. Sage Publications, Thousand Oaks, pp 275–285

Gómez del P, Luis J. (2010) Report of the working group on the use of mercenaries as a means of violating human rights and impending the exercise of the right of peoples to self-determination. UN general assembly, human rights council, fifteenth session, agenda item 3 (a/HRC/15/25), 2 July http://www2.ohchr.org/english/issues/mercenaries/docs/A.HRC.15.25.pdf

Denning D (2000) Cyberterrorism: testimony before the special oversight panel on terrorism, committee on armed services, U.S. house of representatives, 23 May. http://www.stealth-iss.com/documents/pdf/CYBERTERRORISM.pdf

Denyer S (2016) China wants to give all of its citizens a score – and their rating could affect every area of their lives. The Independent, 22 October. http://www.independent.co.uk/news/world/asia/china-surveillance-big-data-score-censorship-a7375221.html?utm_campaign=Echobox&utm_medium=Social&utm_source=Facebook

Department for Homeland Security (2006) Report to the public concerning the multistate anti-terrorism information exchange (MATRIX) pilot project. Privacy Office, U.S. Department of Homeland Security, Washington, DC December. https://www.dhs.gov/sites/default/files/publications/privacy-matrix-122006.pdf

Department for Transport (2016a) Security scanners code of practice. London: Department for Transport. *Gov.UK Guidance: Security scanners implementation information*, Published 22 March 2010. Last updated: 20 December 2016. https://www.gov.uk/government/uploads/system/uploads/attachment_data/file/573478/security-scanner-code-of-practice-2016.pdf

Department for Transport (2016b) Direction to certain aerodrome managers under the aviation security act 1982 Relating to security scanners 2016 (Security Scanners Directive 2016). Gov. UK Guidance: security scanners implementation information. Published: 22 March 2010. Last updated: 20 December 2016. https://www.gov.uk/government/uploads/system/uploads/attachment_data/file/573488/security-scanner-direction-2016.pdf

Dillon M (1996) 25 years of terror: IRA's war against the British. Bantam Books, London

Dodd V (2017) Met police to use facial recognition software at Notting Hill carnival. The Guardian, 5 August. https://www.theguardian.com/uk-news/2017/aug/05/met-police-facial-recognition-software-notting-hill-carnival

Dodd V (2018) Teach five-year-olds about dangers of knife crime, says police chief. The Guardian, 16 January. https://www.theguardian.com/uk-news/2018/jan/15/teach-five-year-olds-about-dangers-of-knife-says-police-chief

Doll N (2016) Experten kritisieren unzureichende Schutzmassnahmen. Welt, Wirtschaft, 21 December. https://www.welt.de/wirtschaft/article160476014/Experten-kritisieren-unzureichende-Schutzmechanismen.html

Dolnik A (2009) Understanding terrorist innovation: technology, tactics, and global trends. Routledge, London/New York

Dymek M, Zackariasson P (eds) (2017) The business of gamification: a critical analysis. Routledge, London/New York

Eckersley, P et al (2018) The malicious use of artificial intelligence: forecasting, prevention, and mitigation. Electronic frontier foundation, 20 February. https://www.eff.org/deeplinks/2018/02/malicious-use-artificial-intelligence-forecasting-prevention-and-mitigation

Enzensberger HM (1979) Der Sonnenstaat des Doktor Herold. Der Spiegel, 18 June. http://www.spiegel.de/spiegel/print/d-40349651.html

Eubulletin (2017) Shielding European cities from terrorism: EU urges protection of 'soft targets'. EUBulletin, 19 October. http://www.eubulletin.com/7691-shielding-european-cities-from-terrorism-eu-urges-protection-of-soft-targets.html

EUR-Lex (2010) Communication from the Commission of the European Parliament and the council on the use of security scanners at EU airports. EUR-Lex access to European Union law, document 52010DC0311. Brussels, 15 June. http://eur-lex.europa.eu/legal-content/EN/TXT/?uri=celex%3A52010DC0311

EUR-Lex (2016) Regulation (EU) 2016/679 of the European Parliament and of the council of 27 April 2016 on the protection of natural persons with regard to the processing of personal data and on the free movement of such data, and repealing directive 95/46/EC (General Data Protection Regulation). Brussels, 27 April. http://eur-lex.europa.eu/eli/reg/2016/679/oj

Evangelista M (2002) The Chechen wars: will Russia go the way of the soviet union? Brookings Institution Press, Washington, DC

Fahrmeir A (2007) Citizenship. The rise and fall of a modern concept. Yale University Press, New Haven/London

FBI (n.d.) Amerithrax or Anthrax investigation. In: History: famous cases & criminals. Federal Bureau of Investigation, Washington, D.C https://www.fbi.gov/history/famous-cases/amerithrax-or-anthrax-investigation

Feinstein JS, Kaplan EH (2011) Counterterror intelligence operations and terror attacks. Public Choice, 149 (December), 281–295. https://link.springer.com/content/pdf/10.1007%2Fs11127-011-9842-7.pdf

Fienberg SE (2008) Homeland insecurity: data mining, privacy, disclosure limitation, and the hunt for terrorists. In: Chen H et al (eds) In terrorism informatics. Knowledge management and data Mining for Homeland Security. Springer, New York, pp 197–218

Findlay J (2017) Offensive behaviour at football act: the demise of the Government's most despised and dangerous legislative folly. Sunday Herald, 25 June

Fisher KD, Creekmore AT III (2014) Making ancient cities: new perspectives on the production of urban places. In: Greekmore AT III, Fisher KD (eds) Making ancient cities: space and place in early urban societies. Cambridge University Press, New York, pp 1–31

Flaherty C (2015) The role of CCTV in terrorist TTPs: camera system avoidance and targeting. Small Wars Journal, November 9

Fleming M (1980) Propaganda by the deed: terrorism and anarchist theory in late nineteenth-century Europe. Studies in Conflict & Terrorism 4(1–4):1–23 https://www.tandfonline.com/doi/pdf/10.1080/10576108008435483

Flusty S (1994) Building paranoia: the proliferation of interdictory space and the Erosion of spatial justice. Forum for Architecture and Urban Design, Los Angeles

Forster K (2017) London terror attack: security barriers installed on three bridges in capital. Independent, 5 June. http://www.independent.co.uk/news/uk/home-news/london-terror-attack-security-barriers-installed-three-bridges-waterloo-lambeth-westminster-a7773476.html

Franklin B (1755) Pennsylvania assembly: reply to the governor. November 11. http://franklinpapers.org/franklin/framedVolumes.jsp?vol=6&page=238a

Franzius C (2015) Das Recht auf informationelle Selbstbestimmung. Zeitschrift für das juristische Studium 3:259–270 http://www.zjs-online.com/dat/artikel/2015_3_906.pdf

Fritz RD (2007) Dean death investigation. In: State attorney forSt Mary's county report. Court House, Leonardtown at http://www.ejfi.org/PDF/Dean_Death_Investigation.pdf

Gandy OH (2002) Data mining and surveillance in the post-9.11 environment. Presentation to the political economy section, IAMCR. Barcelona, July. http://web.asc.upenn.edu/usr/ogandy/iamcrdatamining.pdf

Ganor B (2005) The counter-terrorism puzzle: a guide for decision makers. Transaction, New Brunswick/London

Gapper J (2018) Cambridge Analytica exploited Facebook data with style. Financial Times, 21 March. https://www.ft.com/content/bb24f946-2918-11e8-b27e-cc62a39d57a0

Gardham D (2012) 7/7 bombers planned attack on Bank of England. The telegraph, 30 April. https://www.telegraph.co.uk/news/uknews/terrorism-in-the-uk/9237355/77-bombers-planned-attack-on-Bank-of-England.html

Gellman B and Poitras L (2013) U.S., British intelligence mining data from nine U.S. internet companies in broad secret program. Washington Post, 7 June. https://www.washingtonpost.com/investigations/us-intelligence-mining-data-from-nine-us-internet-companies-in-broad-secret-program/2013/06/06/3a0c0da8-cebf-11e2-8845-d970ccb04497_story.html?utm_term=.7ac02725b7a2

Gibbs S (2016) How safe is voice recognition and fingerprint ID? The Guardian, 19 February. https://www.theguardian.com/technology/2016/feb/19/how-safe-is-voice-recognition-fingerprint-id-hsbc

Golan G (2006) Closing the gateways of democracy: cities and the militarization of protest policing," Cambridge, MA: Massachusetts Institute of Technology

Goodman M (2015) Future crimes. Inside the digital underground and the battle for our connected world. Corgi Books, London

Gottmann J (1989) Since megalopolis. The urban writings of Jean Gottmann. Johns Hopkins University Press, Baltimore/London

Graham S (ed) (2004) Cities, war, and terrorism. Towards an urban geopolitics. Blackwell, Malden/Oxford

Graham S (2010) Cities under siege. The new military urbanism. Verso, London/New York

Green C (2018) Tagged criminals 'could be barred from entire cities'. i Weekend, 24 February

Greenslade R (2014) Police use of Ripa law threatens our civil liberties and press freedom. The Guardian, 01 October. https://ww.theguardian.com/media/greenslade/2014/oct/01/police-press-freedom

Greenwald G, MacAskill E (2013). NSA prism program taps in to user data of apple, google and others. The Guardian, 7 June. https://www.theguardian.com/world/2013/jun/06/us-tech-giants-nsa-data

Greenwald G (2014) The militarization of U.S. police: finally dragged into the light by the horrors of Ferguson, The Intercept, 14 August, at https://theintercept.com/2014/08/14/militarization-u-s-police-dragged-light-horrors-ferguson/

Gross F (1999) Citizenship and ethnicity: the growth and development of a modern multiethnic institution. Greenwood Publishing Group, Santa Barbara

Gulliver (2010) How much do full-body scanners show? The Economist, 8 May. http://www.economist.com/blogs/gulliver/2010/05/full-body_scanners

Gunster S (2004) 'You belong outside': advertising, nature, and the SUV. Ethics Environ 9(2):4–32 (Fall/Winter)

Guyoncourt S, Butterworth B (2018) Firm suspends CEO after undercover exposé. The i Paper, 21 March

Halberstadt H (1994) SWAT team: police special weapons and tactics. Motorbooks International, Osceola

Harding G (2004) X-ray scatter tomography for explosives detection. Radiat Phys Chem 71:869–881 http://www.sciencedirect.com/science/article/pii/S0969806X04003081

Hashiyada M (2011) DNA biometrics. In: Yang J (ed) Biometrics. Intech Open Science https://www.intechopen.com/books/biometrics/dna-biometrics

Hatton C (2015) China 'social credit': Beijing sets up huge system. BBC News China, 26 October. http://www.bbc.co.uk/news/world-asia-china-34592186

Higgens D (2018) New mobile fingerprint device lets police identify suspects in less than a minute. The Independent, 10 February. http://www.independent.co.uk/life-style/gadgets-and-tech/fingerprint-police-identify-suspects-quickly-police-technology-latest-a8203991.html?utm_term=Autofeed&utm_campaign=Echobox&utm_medium=Social&utm_source=Facebook#link_time=1518260586

Homer-Dixon T (2009) The rise of complex terrorism. Foreign Policy. 16 November. http://foreignpolicy.com/2009/11/16/the-rise-of-complex-terrorism/

HSPD-7 (2003) Homeland security presidential directive 7, 17 December, at http://georgewbush-whitehouse.archives.gov/news/releases/2003/12/20031217-5.html

Hu X et al (2008) Text mining the biomedical literature for identification of potential virus/bacterium as bio-terrorism weapons. In: Chen H et al (eds) Terrorism informatics. Knowledge management and data Mining for Homeland Security. Springer, New York, pp 385–406

Hunt AR, Kellerman KF (2000) Development of an analysis tool for performing civil aviation security risk assessment. http://www.akelainc.com/pdf_files/AirportSASSy.pdf

IAEA (2002) Inadequate control of World's radioactive sources. Press Release. Vienna, International Atomic Energy Agency, 24 June. https://www.iaea.org/newscenter/pressreleases/inadequate-control-worlds-radioactive-sources

IDC Herzliya (2015) The Paris Black Friday 13/11/2015 Attacks – What do we know? What should we do? Special Report, 15 November. https://www.ict.org.il/UserFiles/ICT-The-Black-Friday-Paris-Attacks-Nov15.pdf

Introna L, Wood D (2004) Picturing algorithmic surveillance: the politics of facial recognition systems. Surveillance & Society 2(2/3):177–198 http://www.ssoar.info/ssoar/bitstream/handle/document/20067/ssoar-surveillance-2004-23-introna_et_al-picturing_algorithmic_surveillance_the_politics.pdf?sequence=1

Jenkins BM (2006) The new age of terrorism. In: Kamien DG (ed) The McGraw-hill homeland security handbook. McGraw-Hill, New York et al, pp 117–130

Jensen RB (2001) The United States, international policing and the war against anarchist terrorism, 1900–1914. Terrorism and Political Violence, 13(1): 15–46 (Spring)

Johnston PD and Pickering WL (2008) Critical infrastructure protection: a risk management approach. Frontline Security, Summer, pp. 14–16

Johnston AH, Weiss GM (2015) Smartwatch-based biometric gait recognition. In: IEEE Xplore Digital Library. http://ieeexplore.ieee.org/document/7358794/

Jones K (2004) Trust and terror. In: DesAutels P, Walker MU (eds) Moral psychology: feminist ethics and social theory. Rowman & Littlefield, Lanham, MD, pp 3–18

Julian HL (2010) Israeli security expert to Canada: 'full body scanners useless'. Israelnationalnews, 25 April. http://www.israelnationalnews.com/News/News.aspx/137184

Kaarthikeyan SDR (2005) Root causes of terrorism? A case study of the Tamil insurgency and the LTTE. In: Bjørgo T (ed) Root causes of terrorism: myths, realities and ways forward. Routledge, London/New York, pp 131–140

Kaplan R (2000) The coming anarchy: shattering the dreams of the post cold war. Vintage Books, New York

Kaplan R (2013) The revenge of geography. What the map tells us about coming conflicts and the battle against fate. Random House, New York

Kaufman M (2017) German government tells parents to destroy children's smart watches. Mashable, 17 November. http://mashable.com/2017/11/17/german-regulator-tells-people-to-destroy-childrens-smartwatches/?utm_cid=mash-com-fb-main-link#4RPI4gVYbOq5

Kellman B (2007) Bioviolence: preventing biological terror and crime. Cambridge University Press, Cambridge et al.

Khosrokhavar F (2009) Inside Jihadism: understanding jihadi movements worldwide. Routledge (Kindle Edition), London/New York

Kilcullen D (2015) Out of the mountains: the coming age of the urban guerrilla. Hurst (paperback ed.), London

Kilcullen D (2016) Blood year: Islamic state and the failures of the war on terror. Hurst, London

Klitou D (2008) Backscatter body scanners – a strip search by other means. Computer Law & Security Report 24:316–325 http://www.sciencedirect.com/science/article/pii/S0267364908000708

Krahmann E (2010) States, citizens and the privatization of security. Cambridge University Press, Cambridge

Kraska PB (2007) Militarization and policing – its relevance to 21st century policing. Policing:1–13 http://cjmasters.eku.edu/sites/cjmasters.eku.edu/files/21stmilitarization.pdf

Kreissl R (1981) Die präventive Polizei: Auf dem Weg zur gläsernen Gesellschaft? Kritische Justiz 2(81):128–139 http://www.kritv.nomos.de/fileadmin/kj/doc/1981/19812Kreissl_S_128.pdf

Kurz RW, Bartels CK (2007) Chechen suicide bombers. J Slav Mil Stud 20:529–547 www.dtic.mil/dtic/tr/fulltext/u2/a477845.pdf

Kurzweil R (1999) The age of spiritual machines: when computers exceed human intelligence. Viking Press, New York

Lamb H (2017) Facial recognition technology 'fails' at Notting Hill carnival. E&T Engineering and technology, 1 September. https://eandt.theiet.org/content/articles/2017/08/facial-recognition-technology-fails-at-notting-hill-carnival/

Laqueur W (1987) The age of terrorism. Weidenfeld and Nicolson, London

Latour B (2005) Reassembling the social: an introduction to actor-network-theory. Oxford University Press (paperback ed.), Oxford et al.

Lee N (2014) Facebook nation: Total information awareness. Springer, New York et al.

Lee J (2017) Dubai to test ObjectTech-developed biometric border system. Biometric Update, 11 June. http://www.biometricupdate.com/201706/dubai-to-test-objecttech-developed-biometric-border-control-system

Lehr P (2016) (No) princes of the sea: reflections on maritime terrorism. In: Krause J, Bruns S (eds) Routledge handbook of naval strategy and security. Routledge, London/New York, pp 202–214

Leisch H (2002) Gated communities in Indonesia. Cities 19(5):341–350

Levi N (2014) TSA's 'explosive trace detection' needs a dramatic overhaul, Travelers united, 8 September. https://www.travelersunited.org/getting-there/tsas-explosive-trace-detection-needs-a-dramatic-overhaul/

Lewis B (1967) The assassins. A radical sect in Islam. Basic Books, New York

Lewis R (2017) Under surveillance: being watched in modern America. University of Texas Press, Austin

Library of Congress (n.d.) Qaddafi, Muammar. Library of Congress Name Authority File http://id.loc.gov/authorities/names/n81068638.html

Linker KL (2014) Explosives detection personnel portals. In: Yinon J (ed) Counterterrorist detection techniques of explosives. Elsevier, Amsterdam, pp 367–393

Lippold KS (2012) Front burner. Al Qaeda's attack on the USS Cole. Public Affairs, New York

LIST (2017) Using virtual and augmented reality to fight terrorism and crime. *News List*. Luxembourg Institute of Science and Technology. 25 October. https://www.list.lu/en/news/using-virtual-and-augmented-reality-to-fight-terrorism-and-crime/

Lister CR (2015) The Syrian Jihad: Al-Qaeda, the Islamic state and the evolution of an insurgency. Oxford University Press (e-book), Oxford/New York

Lobe A (2017) Wenn Überwachung Spaß macht. Sueddeutsche Zeitung, 4 December. http://www.sueddeutsche.de/kultur/ueberwachung-wenn-ueberwachung-spass-macht-1.3776269

Loo B (2005) The military and counter-terrorism. In: *IDSS Commentaries,* 8 December

LRAD Company (2017) Fact sheet.https://www.lradx.com/about/lrad-public-safety-applications-fact-sheet/

Lyon D (2004) Technology vs. 'terrorism': circuits of City surveillance since September 11, 2001. In: Graham S (ed) Cities, war, and terrorism. Towards an urban geopolitics. Blackwell, Malden/Oxford, pp 297–311

MacAskill E et al (2013) GCHQ taps fibre-optic cables for secret access to world's communications. The Guardian, 21 June. https://www.theguardian.com/uk/2013/jun/21/gchq-cables-secret-world-communications-nsa

Macmillan Buzzword Dictionary (n.d.) Augmented reality. In: Macmillan dictionary, buzzword http://www.macmillandictionary.com/buzzword/entries/augmented-reality.html

Macwirther I (2018) Facebook, Google and Amazon are the robber barons of the modern age . . . break up their monopolies. Sunday Herald, 25 March. http://www.heraldscotland.com/news/16114620.Iain_Macwhirter__Facebook__Google_and_Amazon_are_the_robber_barons_of_the_modern_age/

Madison J (1798) From James Madison to Thomas Jefferson, 13 May 1798. https://founders.archives.gov/documents/Madison/01-17-02-0088

Maguire M, Frois C, eds NZ (2014) The anthropology of security: perspectives from the frontline of policing, counter-terrorism and border control. Pluto Press (paperback), London

Malčik D, Drahansky M (2012) Anatomy of biometric passports. J Biomed Biotechnol:490362. https://www.hindawi.com/journals/bmri/2012/490362/

Mance H (2016) Britain has had enough of experts, says Gove. Financial Times, 3 June. https://www.ft.com/content/3be49734-29cb-11e6-83e4-abc22d5d108c

Manchester Airport (2017) Body scanners at Manchester airport. www.manchesterairport.co.uk/guides-to-travelling/security-and-fasttrack/body-scanners/

Marcuse P (2004) The 'war on terrorism' and life in cities after September 11, 2001. In: Graham S (ed) Cities, war, and terrorism. Towards an urban geopolitics. Blackwell, Malden/Oxford, pp 263–275

Marighella C (1969) Minimanual of the urban Guerrilla. https://www.marxists.org/archive/marighellacarlos/1969/06/minimanual-urban-guerrilla/

Mariner J (2005) The unjust detention of Jose Padilla." FindLaw, 14 September. http://supreme.findlaw.com/legal-commentary/the-unjust-detention-of-jose-padilla.html

Marino D (2018) FakeApp: groundbreaking or dangerous? Artefact Magazine, 13 February. http://www.artefactmagazine.com/2018/02/13/fakeapp-groundbreaking-or-dangerous/

Mascolo G et al (2017) Kaum ein Hacker will zum BND. Sueddeutsche Zeitung, 3 April. http://www.sueddeutsche.de/digital/geheimdienst-kaum-ein-hacker-will-zum-bnd-1.3446843

Mason JE et al (2016) Machine learning techniques for gait biometric recognition. Springer International Publishing, Cham

Matzig G (2018) Ausgeschlossen von der Urbanität. Sueddeutsche Zeitung, 27 February. http://www.sueddeutsche.de/kultur/diesel-fahrverbote-ausgeschlossen-von-der-urbanitaet-1.3883626

McEwen A, Cassimally H (2014) Designing the internet of things. John Wiley & Sons, London

McGoogan C (2017) The end of passport gates? Dubai to test 'invisible' airport checks using facial recognition. The Telegraph, 13 June. http://www.telegraph.co.uk/technology/2017/06/13/end-passport-gates-dubai-test-invisible-airport-checks-using/

Mehra T (2016) Foreign fighters: trends, dynamics and policy responses. ICCT Report, December. https://icct.nl/wp-content/uploads/2016/12/ICCT-Mehra-FTF-Dec2016-2.pdf

Mill JS (1987) Utilitarianism. Penguin (first published 1863), London

Miller CC (2013) Google searches for style. The New York Times, 20 February. http://www.nytimes.com/2013/02/21/technology/google-looks-to-make-its-computer-glasses-stylish.html

Miller N (2016) Brussels attacks: man in the white coat 'blown away' by other bombs, The Sydney Morning Herald, 26 March, at http://www.smh.com.au/world/brussels-attacks-man-in-the-white-coat-blown-away-by-other-bombs-20160326-gnrihw.html

Mills MP, Huber PW (2002) How technology will defeat terrorism: at home and abroad, digital wizardry will keep us safe. City Journal, Winter

Milmo C (2016) It is a matter of when, not if, we are attacked again, warns met police chief. i Paper, 01 August

Minton A (2017) Big capital: who is London for? Penguin, London

Mitchell D (2005) The S.U.V. Model of citizenship: floating bubbles, buffer zones, and the rise of the 'purely atomic' individual. Polit Geogr 24:77–100

Modi S (2011) Biometrics in identity management: concepts to applications. Artech House, Boston/London

de Moncan P (2012) Le Paris d'Haussmann. Les Editions de Mécène, Paris (paperback)

Moon M (2016) New York will use facial recognition to catch terrorists. Engadget, 10 July. https://www.engadget.com/2016/10/07/new-york-crossings-project/

Mueller JE (2006) Overblown: how politicians and the terrorist industry inflate national security threats, and why we believe them. Free Press, New York

Mumford L (1989) The City in history: its origins, its transformations, and its prospects. MJF Books, New York

Munzinger P (2016) Auch das umfassendste Sicherheitskonzept wird solche Anschläge nie ausschließen können, Sueddeutsche Zeitung, 19 July. http://www.sueddeutsche.de/bayern/attacke-mit-axt-und-messer-herrmann-taeter-von-wuerzburg-koennte-sich-selbst-radikalisiert-haben-1.3084851

Myers L (2004) An exploration of voice biometrics. SANS Institute InfoSec Reading Room, 19 April. https://www.sans.org/reading-room/whitepapers/authentication/exploration-voice-biometrics-1436

Naumann M, Bernt M (2009) "When the tap stays dry: water networks in eastern Germany." *Local Environment*. The International Journal of Justice and Sustainability 14(5):461–471

Nelson, Bill et al. 1999. *Cyberterror. Prospects and Implications.* Monterey, CA: Center for the Study of terrorism and irregular warfare. http://www.nps.edu/academics/centers/ctiw/files/Cyberterror%20Prospects%20and%20Implications.pdf

Newman O (1972) Defensible space. People and Design in the Violent City. Architectural Press, London

NIPP (2009) National Infrastructure Protection Plan: partnering to enhance protection and resilience. Department of Homeland Security, Washington D.C. https://www.dhs.gov/xlibrary/assets/NIPP_Plan.pdf

Norris C, Armstrong G (1999) The maximum surveillance society: the rise of CCTV. Berg, Oxford/New York

Osborne S (2015) China has made obedience to the state a game. The Independent, 22 December. https://www.independent.co.uk/news/world/asia/china-has-made-obedience-to-the-state-a-game-a6783841.html

Orwell G (2000) Nineteen eighty-four. Penguin Classics, London (first published 1949)

Pape RA (2005) Dying to win: the strategic logic of suicide terrorism. Random House, New York

Patnaik P (2011) India is rolling out identity cards. The Guardian, 2 March, at https://www.theguardian.com/global-development/poverty-matters/2011/mar/02/india-identity-cards-uidai-debate-poor

PCCIP (1997) Critical foundations: protecting America's infrastructures. President Clinton's commission on critical infrastructure protection, Washington, DC. at http://www.fas.org/sgp/library/pccip.pdf

Peachey P (2016) Police 'used terror powers to spy on officers blowing whistle on racism'. The Independent, 03 January. http://www.independent.co.uk/news/uk/crime/police-accused-of-using-terror-powers-to-spy-on-officers-blowing-whistle-on-racism-a6795036.html

Pedahzur A (2005) Suicide Terrorism. Polity Press, Cambridge

Peterson B (2016) How airport security has changed since 9/11. Condé Nast Traveller, 10 September. https://www.cntraveler.com/story/how-airport-security-has-changed-since-september-11

Prantl H (2010) Horst Herold: Gefangen in den Rastern des Staates. Süddeutsche Zeitung, 17 October. http://www.sueddeutsche.de/kultur/577/404356/text/

Prantl H (2017) Die bleibende Last der RAF-Zeit. Süddeutsche Zeitung, 5 September. http://www.sueddeutsche.de/politik/raf-bleibende-last-der-bleiernen-zeit-1.3647702

Pratt D (2018) Cambridge Analytica's world wide web: a story of data, democracy and dirty tricks. Sunday herald, 25 March. http://www.heraldscotland.com/news/16114621.Cambridge_Analytica__39_s_world_wide_web__a_story_of_data__democracy_and_dirty_tricks/

Price JC, Forrest JS (2016) Practical aviation security. Predicting and preventing future threats. Amsterdam et al. Butterworth-Heinemann (3rd edition)

Quilliam (n.d.) About Us, at http://www.quilliamfoundation.org/about/

Rapoport D (2004) The four waves of modern terrorism. In: Cronin AK, Ludes JL (eds) Attacking terrorism. Georgetown University Press, Washington, DC, pp 46–73

Rathgeb C, Busch C (2012) Multi-biometric template protection: issues and challenges. In: Yang J (ed) New trends and developments in biometrics. InTech, pp 173–190 https://www.intechopen.com/books/new-trends-and-developments-in-biometrics/multi-biometric-template-protection-issues-and-challenges

Raytheon (2009) NATO Naval Armaments Group. Workshop on Counter Piracy Equipment and Technologies. Raytheon, 4 June. https://web.archive.org/web/20120524034612/http://www.nato.int/structur/AC/141/pdf/S-B/Raytheon.pdf

Reed A, Pohl J, Jegerings M (2017) The four dimensions of the foreign fighter threat: making sense of an evolving phenomenon. ICCT Policy Brief (June). https://icct.nl/wpcontent/uploads/2017/06/ICCT-Reed-Pohl-The-Four-Dimensions-of-the-Foreign-Fighters-Threat-June-2017.pdf

Reid J (2007) Securing the critical National Infrastructure. RUSI Journal 152(3):14–17

Relyea HC and Seifert JW (2005) Information sharing for homeland security: a brief overview. CRS Report for Congress, updated January 10. https://fas.org/sgp/crs/RL32597.pdf

Ressam A (2002) Testimony of Ahmad Ressam (July 3–5, 2001). In Anti-American terrorism and the Middle East: a documentary reader, ed. Barry Rubin and Judith Colp Rubin, 173. Oxford et al: Oxford University Press

Revell T (2016) Dutch police use augmented reality to investigate crime scenes. New Scientist, 21 November. https://www.newscientist.com/article/2113450-dutch-police-use-augmented-reality-to-investigate-crime-scenes/

Ricks TE (2007) Fiasco: The American military adventure in Iraq, 2003–2005. Penguin, London/New York

Riedel B (2012) Mumbai attacks: four years later. Brookings Op-Ed, 26 November. https://www.brookings.edu/opinions/mumbai-attacks-four-years-later/

Rifkind H (2018) Facebook and friends assume we are fools. The Times, 19 March. https://www.thetimes.co.uk/article/facebook-and-friends-assume-we-are-fools-3l6zz99nc

Robb J (2007) Brave new war. The next stage of terrorism and the end of globalization. John Wiley and Sons, Hoboken

Robertson N, Cruikshank P, Lister T (2012) Documents give new details on al Qaeda's London bombings CNN, 30 April. https://edition.cnn.com/2012/04/30/world/al-qaeda-documents-londonbombings/index.html

Robinson M (2016) This app lets you make easy money by selling personal data to marketers. Business Insider, 10 June. http://uk.businessinsider.com/datawallet-lets-you-sell-your-data-2016-6?r=US&IR=T

Roper K (1999) Risk Management for Security Professionals. Rothstein Associates, Brookfield

Rosenzweig P, Kochems A (2005) Risk assessment and risk management: necessary tools for homeland security, Backgrounder No 1889, October 25. http://www.heritage.org/Research/Reports/2005/10/Risk-Assessment-and-Risk-Management-Necessary-Tools-for-Homeland-Security

Rougerie J (2014) La Commune de 1871. Presses Universitaires de France (mass market paperback), Paris

Rumsfeld D (2002) Briefing to the US Department of defense. Washington, DC.: Department of Defense. February 12. http://archive.defense.gov/Transcripts/Transcript.aspx?TranscriptID=2636

Russell J (2018) Chinese police are using smart glasses to identify potential suspects. TechCrunch, 8 February. https://techcrunch.com/2018/02/08/chinese-police-are-getting-smart-glasses/

Savičić G, Savić S (eds) (2016) Unpleasant design. G.L.O.R.I.A (Kindle edition), Belgrade

Savov V (2017) Google glass gets a second chance in factories, where it's likely to remain. The Verge, 18 July https://www.theverge.com/2017/7/18/15988258/google-glass-2-enterprise-edition-factories

Schleutker-Franke S (2017) Handy-Aufkleber sollen Privatsphäre schützen. Swim, 18 January. https://swim.de/aktuell/handy-aufklebern-privatsphaere-schuetzen-67101

Schneier B (2013) It's smart politics to exaggerate terrorist threats. CNN, May 20. www.edition.cnn.com/2013/05/20/opinion/schneier-security-politics/

Scott-Clark C, Levy A (2013) The siege: 68 hours inside the Taj Hotel. Penguin Books, New York

Seale P (1992) Abu Nidal: a gun for hire. Hutchinson, London et al.

Seiderer B (2004) Horst Herold und das Nürnberger Modell (1966–1971). Eine Fallstudie zur Pionierzeit des Polizeilichen EDV-Einsatzes in der Reformära der Bundesrepublik. In: Mitteilungen des Vereins für Geschichte der Stadt Nürnberg, Bd, vol 91, pp 317–350 http://periodika.digitale-sammlungen.de/mvgn/Blatt_bsb00001092,00382.html

Senesac L, Thundat TG (2008) Nanosensors for trace explosive detection. Mater Today 11(3): 28–36 (March). http://www.sciencedirect.com/science/article/pii/S1369702108700178

Sengupta K (2016) Isis claims responsibility for the Orlando gay nightclub shooting. But that doesn't mean they knew anything. The Independent, 13 June 2016, at http://www.independent.co.uk/voices/isis-claims-responsibility-orlando-gay-nightclub-shooting-florida-attack-omar-mateen-gunman-a7079721.html

Sequin M (2017) JetBlue is the latest to use facial recognition technology in airports. Mashable UK, 03 June. http://mashable.com/2017/06/03/facial-recognition-airports-boarding-pass/#HVghHPRcSqq5

Sharma RC et al (2013) Portable detection systems for standoff sensing of explosives and hazardous materials. Opt Commun 309:44–49

Sheridan K (2010) 'Tunnel of truth' for air travelers is a long way off. PhysOrg Technology, November 24. https://phys.org/news/2010-11-tunnel-truth-air.html

Simmons D (2017) BBC fools HSBC voice recognition security system. BBC News, 19 May 2017. http://www.bbc.co.uk/news/technology-39965545

Simon JD (2001) The alphabet bomber (1974). In: Tucker JB (ed) Toxic terror. Terrorist use of chemical and biological weapons. MIT Press, Cambridge, MA and London, pp 71–94

Singh S, Singh M (2003) Explosives detection systems (EDS) for aviation security. Signal Process 83:31–55

Sky News (2013) Taser shotgun tests scrapped by Home Office. Sky News, 16 July. https://news.sky.com/story/taser-shotgun-tests-scrapped-by-home-office-10440018

Sloan S (2007) Educating the next generation of counterterrorism professionals. In: Forest JJF (ed) Countering terrorism and insurgency in the 21st century. Praeger Security International, Westport/London, pp 11–32

Snälla B (2017) Explosives detection in security – how it works, and decoding 'alarm resolution'." Snällabolaget Home, at http://snallabolaget.com/?page_id=3098

Solon O (2012) How much data did Facebook have on one man? 1,200 pages of data in 57 categories. Wired, 28 December. http://www.wired.co.uk/article/privacy-versus-facebook

Solove DJ (2004) The digital person: technology and privacy in the information age. New York University Press, New York/London

Sommerlad J (2018) Mozilla and Commerzbank pull Advertising from Facebook over Cambridge Analytica breach. The Independent, 23 March. https://www.independent.co.uk/life-style/gadgets-and-tech/news/facebook-data-breach-latest-mozilla-commerzbank-advertising-cambridge-analytica-mark-zuckerberg-a8269936.html

Sorkin M (2004) Urban warfare: a tour of the battlefield. In: Graham S (ed) Cities, war, and terrorism. Towards an urban geopolitics. Blackwell, Malden/Oxford, pp 251–262

Sparke M (2008) Fast capitalism/slow terror: cushy cosmopolitanism and its extraordinary others. In: Amoore L, de Goode M (eds) Risk and the war on terror. Routledge, London/New York, pp 133–157

Spiegel (1971) Kommissar computer. Der Spiegel, 28 June. http://www.spiegel.de/spiegel/print/d-43243024.html

Spiegel Online (2013) Schnüffelsoftware XKeyscore: Deutsche Geheimdienste setzen US-Spähprogramm ein. Spiegel Online, 20 July. http://www.spiegel.de/politik/deutschland/bnd-und-bfv-setzen-nsa-spaehprogramm-xkeyscore-ein-a-912196.html

Spiegel Online (2018) Rennen und Fallen sind in Mannheim bald verdächtig: Projekt mit intelligenten Kameras. Spiegel Online Netzwelt 15 February. http://www.spiegel.de/netzwelt/netzpolitik/mannheimer-weg-2-0-pilotprojekt-mit-intelligenten-kameras-startet-bald-a-1193622.html

Stiglitz JE, Bilmes L (2008) The three trillion dollar war: the future cost of the Iraq conflict. W. W. Norton, New York

Strittmatter K (2018) Absolute Kontrolle. Sueddeutsche Zeitung, 1 February. http://www.sueddeutsche.de/digital/digitale-ueberwachung-in-china-absolute-kontrolle-1.3849464

Sunday Herald (2018) Muse unveiled: top authors on the books that made them. Sunday Herald, 11 March

Sunstein CR (2014) Nudging: a very short guide. In: Digital access to scholarship at Harvard https://dash.harvard.edu/bitstream/handle/1/16205305/shortguide9_22.pdf?sequence=4

Swindon P (2016) Anti-terror police: 'Scotland is the safest part of the UK'," Sunday Herald, 31 July

Tarantola A (2011) How to terror-proof the new world trade center. Gizmodocom, 9 September. https://gizmodo.com/5838327/terror-proofing-the-new-world-trade-center

Taw JM, Hoffman B (1994) The urbanization of insurgencies: the potential challenge to U.S. Army operations. RAND, Santa Monica

Taylor D (1994) Citizenship and social power. In: Turner BS, Hamilton P (eds) Citizenship: critical concepts, vol 1. Routledge, New York/London, pp 136–147

ThaiTech (2018) Privacy warning as 7-Eleven in Thailand to start scanning customer's [sic] faces. ThaiVisa, 16 March. https://tech.thaivisa.com/privacy-warning-7-eleven-thailand-start-scanning-customers-faces/27775/

The Economist (2015) Blockchains: the great chain of being sure about things. The Economist, 31 October. https://www.economist.com/news/briefing/21677228-technology-behind-bitcoin-lets-people-who-do-not-know-or-trust-each-other-build-dependable

The Economist (2016) Surveillance: halting the hate: a new technique for removing radical propaganda. The Economist, 25 June

The Economist (2017) Even better and cheaper, face-recognition technology is spreading. De Economist, 9 September. https://www.economist.com/news/business/21728654-chinas-megvii-has-used-government-collected-data-lead-sector-ever-better-and-cheaper

Thompson T (2011) Full Body Scanners in Airports: Advantages and Disadvantages. IT Course: Full Body Scanners in Airports, March 6. www.tthompson-fullbodyscannersinairports.blogspot.co.uk

Tibi B (1995) Les conditions d'un 'euro-Islam'. In: Bistolfi R, Zabbal F (eds) Islams d'Europe: intégration ou insertion communautaire? Editions de l'Aube, Paris

Timberg C, Dwoskin E, Hamburger T (2018) Trump campaign consultant took data about millions of users without their knowledge. The Washington Post, 18 March. https://www.washingtonpost.com/business/economy/trump-campaign-consultant-took-data-about-millions-ofusers-without-their-knowledge/2018/03/17/bab541f4-2a18-11e8-ab19-06a445a08c94_story.html?utm_term=.99842c82a80e

Travis A (2017) UK police to lose phone and web data search authorisation powers. The Guardian, 30 November. https://www.theguardian.com/technology/2017/nov/30/police-to-lose-phone-and-web-data-search-authorisation-powers

Wagner AR (2012) Counter-terrorism technologies – taking stock on 9/11. The Huffington Post, November 11. www.huffingtonpost.com/abraham-r-wagner/counterterrorism-technolo_b_1874521.html

Walters W (2008) Putting the migration-security complex in its place. In: Amoore L, de Goode M (eds) Risk and the war on terror. Routledge, London/New York, pp 158–177

Wardrop M (2011) Councils have mounted millions of snooping operations in past decade, finds report. The Telegraph, 04 November. http://www.telegraph.co.uk/news/uknews/crime/8868757/Councils-have-mounted-millions-of-snooping-operations-in-past-decade-finds-report.html

Weizman E (2006) The art of war. Frieze Magazine, May 6. www.frieze.com/article/art-war

Weller C (2017) There's a secret technology in 90 US cities that listens to gunfire 24/7. Business Insider UK, 27 June. http://uk.businessinsider.com/how-shotspotter-works-microphones-detecting-gunshots-2017-6?r=US&IR=T

Wells K, Bradley DA (2012) A review of X-ray explosives detection techniques for checked baggage. Appl Radiat Isot 70:1729–1746 http://www.sciencedirect.com/science/article/pii/S0969804312000127

White N (2011) The privatisation of military and security functions and human rights: comments on the UN working Group's draft convention. Human Rights Law Review 11(1):133–151

Wiberg H (1974) Are urban guerillas possible? In: Niezing J (ed) Urban guerilla. Studies on the theory, strategy and practice of political violence in modern societies. Rotterdam University Press, Rotterdam, pp 11–21

Wiles P (2017) Metropolitan Police's use of Facial Recognition Software at the Notting Hill Carnival, 2017. Gov.UK Press release, 23 August. https://www.gov.uk/government/news/metropolitan-polices-use-of-facial-recognition-technology-at-the-notting-hill-carnival-2017

Williams P, Selle W (eds) (2016) Military contingencies in megacities and sub-megacities. Carlile Barracks. United States Army War College Press (December), PA

Wöhrle C (2017) Ehrenwerte Familien: Libanesische Clans in Deutschland. Focus Online, 24 April 2017. http://www.focus.de/politik/deutschland/kriminalitaet-eine-ehrenwerte-familie_id_6956747.html

Wolfendale J (2007) Terrorism, security, and the threat of counterterrorism. Studies in Conflict & Terrorism 30

Wong JC (2018) Amazon working to fix Alexa after users report random burst of 'creepy' laughter." The Guardian, 7 March. https://www.theguardian.com/technology/2018/mar/07/amazon-alexa-random-creepy-laughter-company-fixing

Woodward JD Jr (2001) Biometrics: facing up to terrorism. RAND Issue Paper, October. Santa Monica: RAND. www.rand.org/content/dam/rand/pubs/issue_papers/2005/IP218.pdf

Yang X (2016) Der Verdacht. SZ Magazin, 16 September. http://www.sueddeutsche.de/digital/gehacktes-handy-der-verdacht-1.3161193

Yang J, Shi Y, Yang J (2012) Finger-vein image restoration based on a biological optical model. In: Yang J, Xie SJ (eds) New trends and developments in biometrics. InTech, pp 59–76 https://www.intechopen.com/books/new-trends-and-developments-in-biometrics/finger-vein-image-restoration-based-on-a-biological-optical-model

Zheng P, Hall RD (2003) Pedestrian guard railing: a review of criteria for installation. Final report to transport of London. Transportation Research Group, University of Southampton, Southampton October. http://content.tfl.gov.uk/pedestrian-guardrailing-review-of-criteria-for-installation.pdf

Zureik E (2006) (with contribution from Karen Hindle). Governance, security and technology: the case of biometrics. Studies in political economy 73 (Spring/Summer), pp. 113–137

Index

P. Lehr, *Counter-Terrorism Technologies*, Advanced Sciences and Technologies for Security Applications, https://doi.org/10.1007/978-3-319-90924-0

The manufacturer's authorised representative in the EU is Springer Nature Customer Service Centre GmbH, Europaplatz 3, 69115 Heidelberg, Germany. If you have any concerns regarding our products, please contact ProductSafety@springernature.com

Printed and bound by CPI Group (UK) Ltd, Croydon, CR0 4YY
25/09/2026
02222857-0007